Robert Pęczkowski

Colour illustrations by
Artur Juszczak

Lockheed P-38J – L
Lightning

STRATUS

Published in Poland in 2013
by STRATUS s.c.
Po. Box 123,
27-600 Sandomierz 1, Poland
e-mail: office@mmpbooks.biz
for
Mushroom Model Publications,
3 Gloucester Close,
Petersfield,
Hampshire GU32 3AX
e-mail: rogerw@mmpbooks.biz

© 2013 Mushroom Model
Publications.
http://www.mmpbooks.biz

**ISBN
978-83-61421-69-6
Second edition**

Editor in chief
Roger Wallsgrove

Editorial Team
Bartłomiej Belcarz
Robert Pęczkowski
Artur Juszczak
Dariusz Karnas
James Kightly

Colour drawings
Artur Juszczak

Scale plans
Dariusz Karnas

3D Drawings
Dariusz Grzywacz

DTP & Layout
Stratus

Printed by
Drukarnia Diecezjalna,
ul. Żeromskiego 4,
27-600 Sandomierz
www.wds.pl
marketing@wds.pl

PRINTED IN POLAND

Table of contents

Acknowledgements

The author would like to thank the late Arthur Lochte,, who had provided much of the material needed to write this book.

Bibliography

Books (selected)
Angelucci, Enzo and Bowers Peter, *The American Fighter*, Orion Books, 1987
Brown Wings, *Captain Eric of the Weird and Wonderful*, Airlife, 1985
Davis, Larry, *P-38 Lightning, Walk Around*, Squadron Publications, 2003
Francillon, Rene J., *Lockheed Aircraft Since 1913*, Naval Institute Press, 1987
Freeman, Roger, *American Eagles, USAAF Colours 2*, Classic Publications, 2001
Green, William, *Famous Fighters of the Second World War*, Doubleday, 1967
Jarski, Adam, *Monografie Lotnicze No 68, 69, 70, P-38 Lightning*, AJ-Press, 2000-2001
Karnas, Dariusz and Strzelczyk, Franciszek, *P-38J/L Modelmania No 8*, AJ-Press, Gdańsk 2007
Kinzey, Bert, *P-38 Lightning Part 2. P-38J Through P-38M*, Squadron Publications, 1998
Lockheed P-38 Lightning, Aero Team Vol 5
Stafford, Gene B, *P-38 In Action no 25*, Squadron Publications, 1976
Swanborough Gordon and Bowers, Peter M., *United States Military Aircraft since 1909*, Smithsonian Institution Press, 1989
War Planes of the Second World War, Fighters, Volume Four, Doubleday, 1964
Weber, Le Roy, *The P-38J-M Lockheed Lightning*, Profile Publications, Profile Publications, Ltd, 1965

Previous page: 1st Lt. Fred C. Eberle of the 333rd FS, 318th FG drew a bit too much attention during a low level run over Iwo Jima on 5 January 1945. He managed to nurse his P-38L, "Ripper" (serial unknown), back to a safe landing on Saipan.

Get in the picture!

Do you have photographs of historical aircraft, airfields in action, or original and unusual stories to tell? MMP would like to hear from you! We welcome previously unpublished material that will help to make MMP books the best of their kind. We will return original photos to you and provide full credit for your images. Contact us before sending us any valuable material: rogerw@mmpbooks.biz

Introduction

At the time when the radical configuration P-38 Lightning entered service, its performance was better than any other aircraft in use with the US Army Air Corps. Its speed, service ceiling, and rate of climb were superior to those of the P-39 Airacobra or P-40 Kittyhawk. Even the arrival of the P-47 Thunderbolt failed to challenge the P-38's supremacy as an excellent escort fighter.

The Lightning was not free from faults, though. Pilots complained about insufficient cockpit heating and icing at high altitudes. Supercharger air cooling was a greater problem. The intercoolers were located in the wing leading edges between the engines and the cockpit, and poor cooling prevented P-38 pilots from using the full power of their Allison V-1710-89/91 engines. The maximum power output was often reduced to 1,250 hp, under front line conditions, instead of 1,425 hp that the engines were capable of. Wing-mounted intercoolers worked satisfactorily up to about 1,000 hp, but above that value they were not able to cope with the cooling requirements, and, additionally, the long intercooler piping was prone to frequent failures.

Moreover, aircraft used in Britain suffered from other engine problems. Recent research indicates that fuel quality issues may have exacerbated a demanding engine, turbosupercharger and altitude problem. Fuel quality was not so critical for engines with mechanical superchargers, or radial engines with turbosuperchargers; but with this in-line engine and turbosupercharger it led to ineffective operation at high altitudes. (It is worth noting that Charles Lindbergh discovered that P-38 pilots were not getting nearly the best out of their engines – using too high RPM in auto rich – in the Pacific theatre.) Even the repositioning of the intercooler from the J model onwards failed to cure the problem and the P-38s were gradually withdrawn from units based in Britain, as Major General Billy Mitchell favoured P-51 Mustangs. By the end of the war only the 474th FG was operational in Europe on Lightnings, though it was well liked in Europe in the interdiction and bomber role.

In the Pacific the situation was quite the reverse. The Lightning was much

P-38H of 80th FS, 8th FG, 5th AAF at Dobodura, New Guinea, September 1943. (T. Kopański coll.)

preferred by USAAF units, in contrast to the alternatives that were on offer and the capabilities of the P-38 were virtually ideal for the conditions in this theatre. Having two engines was a great advantage for long flights over the sea! This aircraft was flown by the two top scoring American pilots of WWII, Bong and McGuire, as well as by many other aces.

General assessment of the Lightning is favourable. After the initial intercooler, poor cockpit heating, and icing problems were solved, after hydraulic assisted ailerons were fitted (this powered aileron system being the first on a major production fighter) and airbrakes were introduced to facilitate dive recovery, the aeroplane was much more combat capable.

Its weak points, that were never cured, included poor rearward visibility, problems with the fit of the wing-fuselage fillet, and a high risk of injury to the pilot when he was forced to bail out, due to the tail arrangement. Like the P-47 Thunderbolt, the turbosuperchargers on the P-38 were challenging to work on, both due to access and their complexity. Compared to aircraft of the previous generation, this was a more

P-38 prototype: the YP-38 at the factory airfield, unveiled to the press.
(Lockheed)

difficult aircraft to maintain. The handed propellers, critical for the combat manoeuvrability of the type, required different reduction gear units for each engine which naturally complicated their maintenance. The aircraft was also neither easy nor cheap to manufacture.

However, the Lightning's good points far outnumbered the shortcomings. The handed propellers, that gave so many headaches to the ground crew, cancelled out the propeller torque that made life so difficult for single-engined fighter pilots. As well as the previously mentioned enhanced combat capability, they gave the Lightning better stability and manoeuvrability.

One of the original objectives for the Lightning's layout was to concentrate the armament. Fitting the machine guns along the fuselage axis, closely grouped together, facilitated aiming and fire concentration regardless of the target's distance, and assisted better gunnery by the pilot. The aircraft's structure was strong and resistant to damage, the twin boom layout giving a slight edge in structural integrity over more conventional layouts. Naturally, the twin engine advantage increased the chances of survival. The Lightning could fly on one engine with no major difficulties, and still hold a decent speed. The sheer size of the twin-engined fighter allowed it to take a sizeable bomb load, in the end up to 4,000lb (not far short of the B-17F's normal 5,000lb). The performance – maximum speed and range – were among the best of contemporary fighters.

The Lightning was developed from the outset as a single-seater, twin-engined air superiority fighter, and this is the main difference with respect to other twin-engined fighters of the time.

The P-38, despite the problems alluded to here, was unarguably one of the best fighters of WWII, notably a rarity in being a twin-engined fighter able to hold its own (in the hands of a competent pilot) against single-engined machines. It also became one of the best reconnaissance aircraft.

Maintenance of a P-38H of 38th FS, based at Nuthamstead, England, October 1943.
(US National Archives)

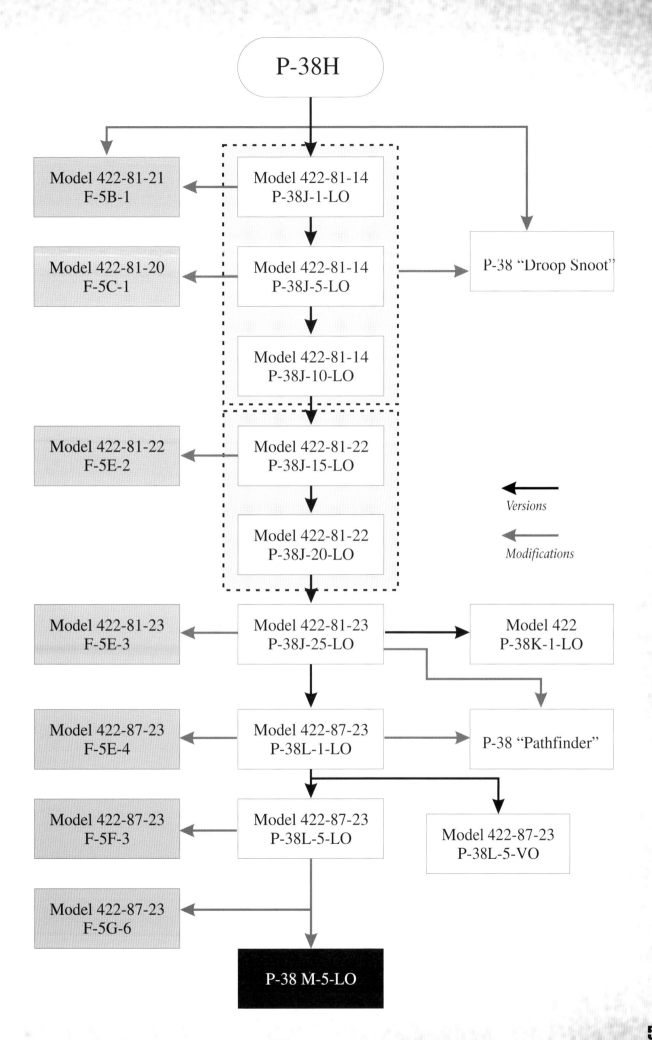

P-38H

Model 422-81-21
F-5B-1

Model 422-81-14
P-38J-1-LO

Model 422-81-20
F-5C-1

Model 422-81-14
P-38J-5-LO

P-38 "Droop Snoot"

Model 422-81-14
P-38J-10-LO

Model 422-81-22
F-5E-2

Model 422-81-22
P-38J-15-LO

Versions

Model 422-81-22
P-38J-20-LO

Modifications

Model 422-81-23
F-5E-3

Model 422-81-23
P-38J-25-LO

Model 422
P-38K-1-LO

Model 422-87-23
F-5E-4

Model 422-87-23
P-38L-1-LO

P-38 "Pathfinder"

Model 422-87-23
F-5F-3

Model 422-87-23
P-38L-5-LO

Model 422-87-23
P-38L-5-VO

Model 422-87-23
F-5G-6

P-38 M-5-LO

5

VERSIONS

P-38J (Model 422)

As previously mentioned, the earlier versions of the P-38 were not free of shortcomings. Lockheed designers developed a new version of the Lightning, taking into account both the requirements of the

P-38J-10-LO, s/n 42-68043, (N2-D), "Aboob" of 383rd FS, 364th FG piloted by Capt. Lee S. Ayoub photographed somewhere in UK, 1944. (US National Archives)

air service, and their own independent research. Particular attention was paid to improved flying characteristics, structural strength, and relieving the pilot from much of the manual engine management. This, the new P-38J, introduced in August 1943, featured a number of improvements:

First of all, new intercoolers were introduced in redesigned (more blunt-looking) engine nacelles. This allowed a significant simplification of the turbosupercharger air cooling system. The space thus gained in the leading edges of the outer wing panels

was used to accommodate additional fuel tanks. The tanks were fitted from version P-38J-5LO on, and all the P-38J-5LO and P-38J-10LO aircraft featured these tanks. This subtype can be recognised by the

additional fuel tank filling points near the leading edge. The additional tanks required the wing to be reinforced by fitting additional sub-spars. In addition, the glycol radiator and turbosupercharger air intakes were reshaped. These were enlarged

and better streamlined to reduce drag. The cockpit was also redesigned.

The first production block, P-38 (Model 422-81-14) covered 1,010 aircraft built in three versions:

P-38J-1-LO – 10 built, test batch;
P-38J-5-LO – 210 built, leading edge tanks introduced;
P-38J-10-LO – 790 built, flat windscreen with integral bullet-proof glass introduced.

The next production block, Model 422-81-22, was built in two versions:

P-38J-15-LO – 1,400 built, with a different electric system (fuses replaced with circuit breakers, for example) and pilot's armour;
P-38L-20-LO – 350 built, turbosupercharger control improved.

Subsequent production block, Model 422-81-23, built in just one version:

P-38-J-25-LO – 210 built.

This version introduced special compressibility flaps that facilitated dive recovery. These were fitted under the wings, so as not to affect the airflow over the tail. They were electric-operated and when deflected, they shifted the centre of pressure of the wing which greatly assisted dive recovery. This version also introduced hydraulic-assisted ailerons, greatly improving lateral control. Before that, at high speeds, P-38 pilots encountered major difficulties with fast aileron control and this restriction was used by enemy pilots to evade Lightnings by making fast rolls. Visually, this modification can be identified by the absence of aileron trim tabs.

P-38J serial numbers	
42-12867—12869	Lockheed P-38J-1-LO
42-13560—13566	Lockheed P-38J-1-LO
42-67102—67311	Lockheed P-38J-5-LO
42-67402—68191	Lockheed P-38J-10-LO
42-103979—104428	Lockheed P-38J-15-LO
43-28248—29047	Lockheed P-38J-15-LO
44-23059—23208	Lockheed P-38J-15-LO
44-23209—23558	Lockheed P-38J-20-LO
44-23559—23768	Lockheed P-38J-25-LO

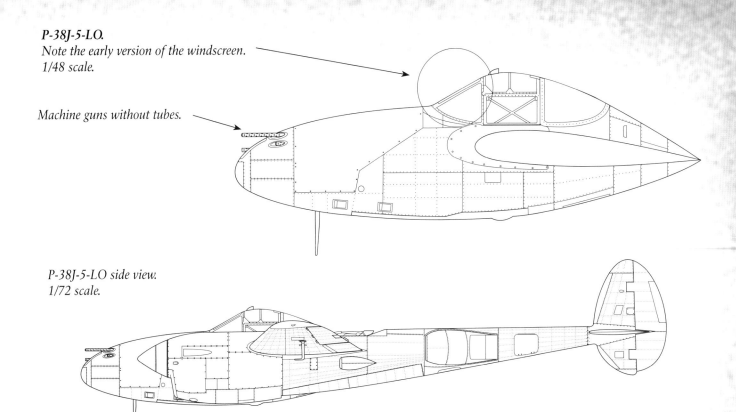

P-38J-5-LO.
Note the early version of the windscreen.
1/48 scale.

Machine guns without tubes.

P-38J-5-LO side view.
1/72 scale.

P-38J-10-LO, s/n 42-67980, (5Y-F -) 384th FS, 364th FG, 8th AF. Assigned to Capt. John C. Ford being recovered after making an emergency landing at Honnington, England on 30 May, 1944. (US National Archives)

71912 A.C.

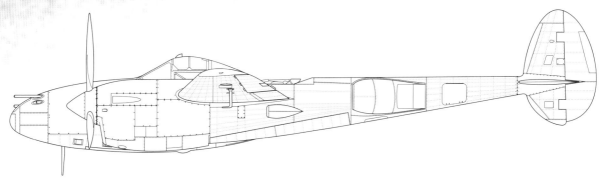

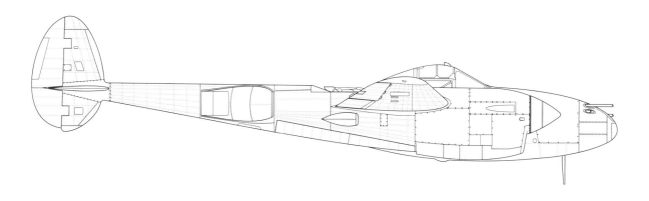

P-38J-15-LO side views. 1/72 scale.

In the foreground is P-38J-10-LO, s/n 42-67811, (CG-H) of 38th FS, 55th FG, 8th AF. The third aircraft it is also P-38J-10-LO, s/n 42-68132 (CG-I) from the same unit. (US National Archives)

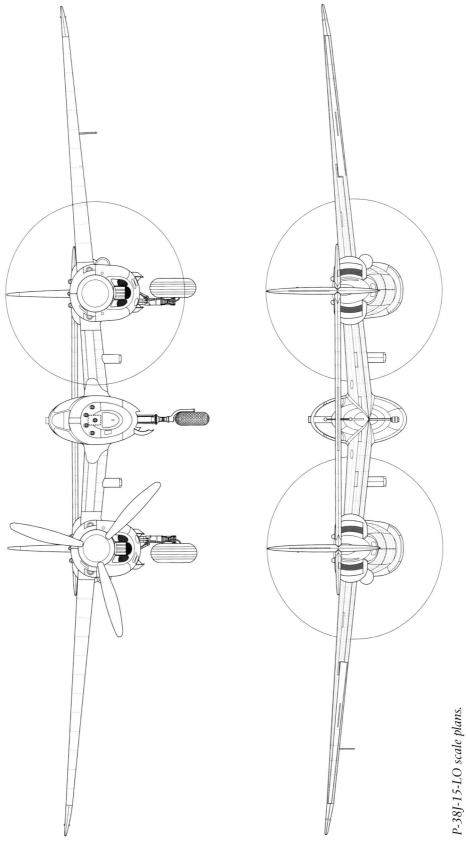

P-38J-15-LO scale plans.
1/72 scale.

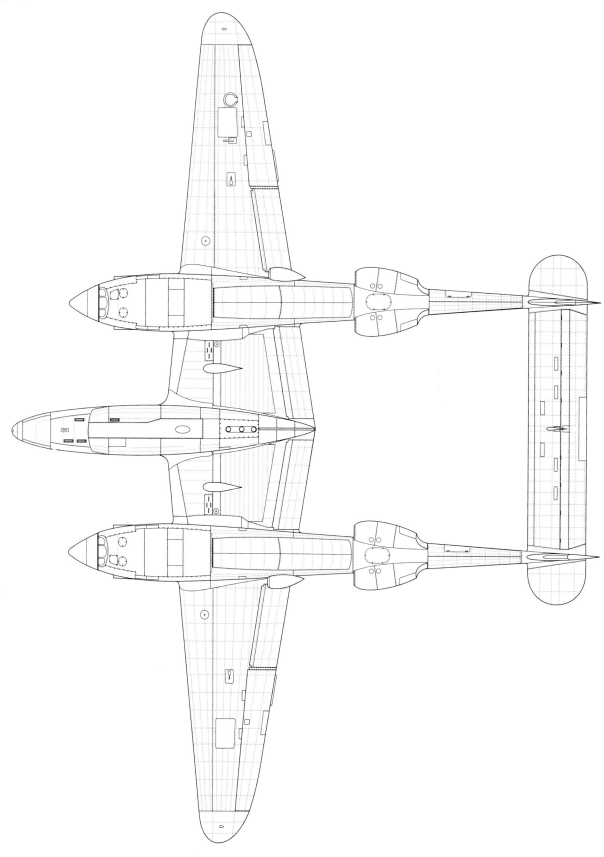

P-38J-15-LO scale plans.
1/72 scale.

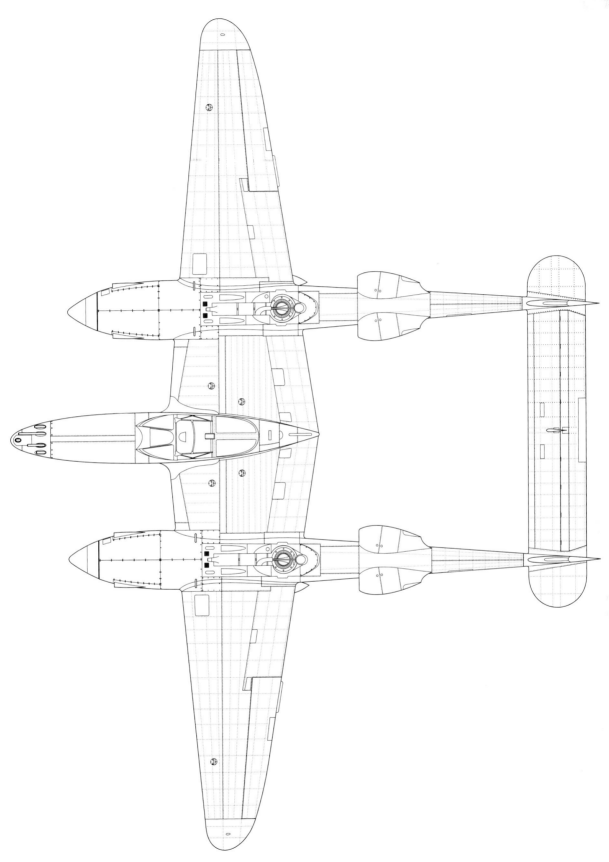

P-38J-15-LO scale plans.
1/72 scale.

P-38J-25-LO side elevation. 1/72 scale.

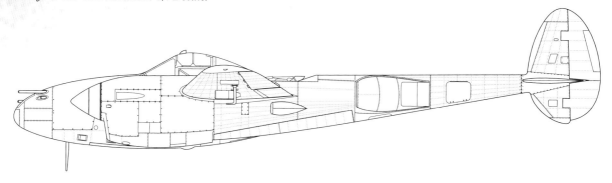

Right: P-38J-10-LO, s/n 42-67816, (CL-P), "Princess Pat" of 338th FFS, 55th FG, 8th AF. Assigned to Maj. Chester A. Patterson. It was shot down (believed to be by AAA) on 10 February 1944 while being flown by 2nd Lt. Jack R. Foster. (US National Archives)

Below: P-38J-10-LO assigned to 1st Lt. William G. Baumeister of the 459th FS "Twin Tailed Dragons", 80th FG. He was officially credited with damaging a Ki-43 Oscar on 23 May 1944. (US National Archives)

54167 A.C

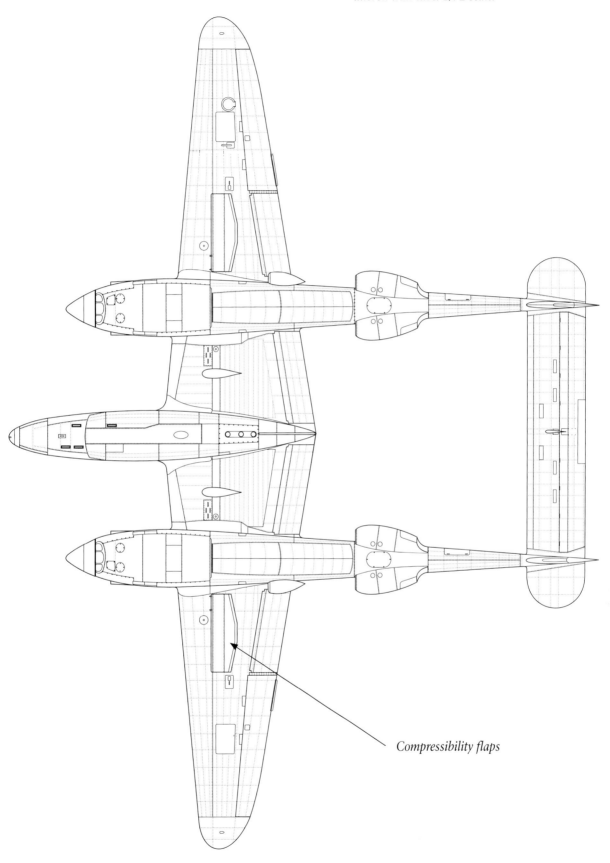

P-38J-25-LO underside view. Note compressibility flaps that facilitated dive recovery and the absence of aileron trim tabs. 1/72 scale.

Compressibility flaps

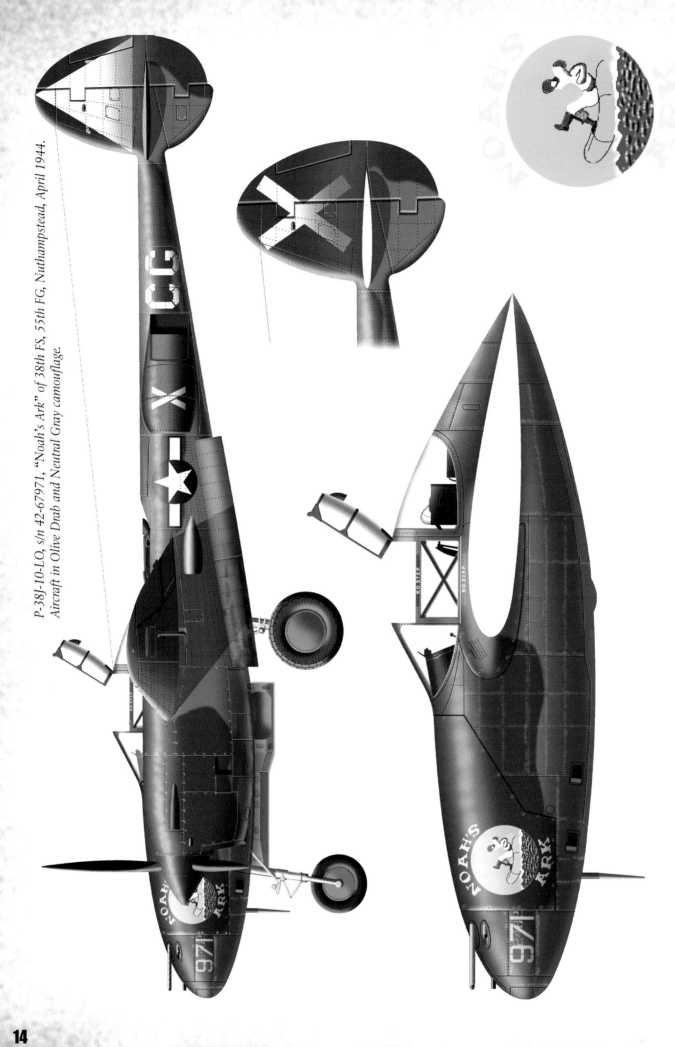

P-38J-10-LO, s/n 42-67971, "Noah's Ark" of 38th FS, 55th FG, Nuthampstead, April 1944. Aircraft in Olive Drab and Neutral Gray camouflage.

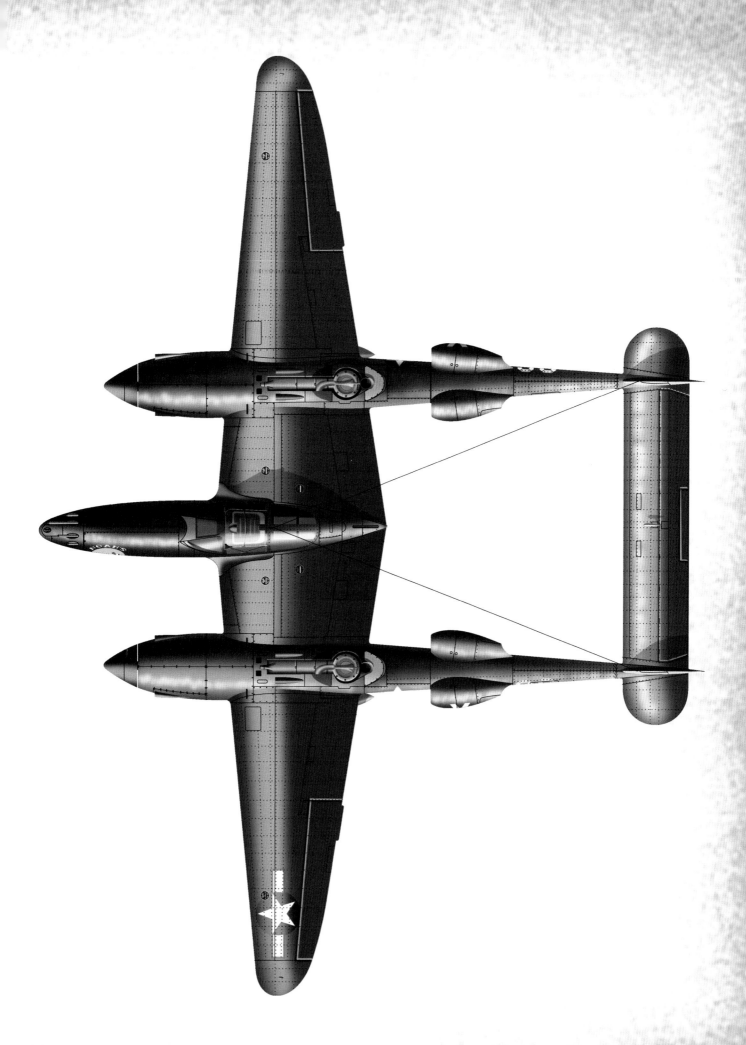

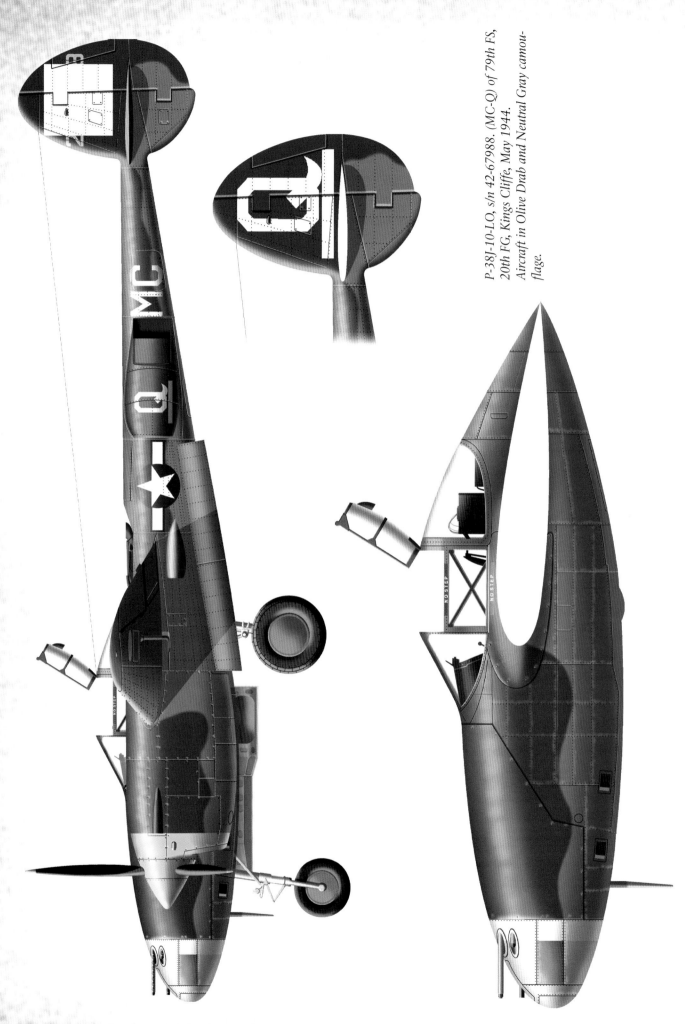

P-38J-10-LO, s/n 42-67988. (MC-Q) of 79th FS, 20th FG, Kings Cliffe, May 1944. Aircraft in Olive Drab and Neutral Gray camouflage.

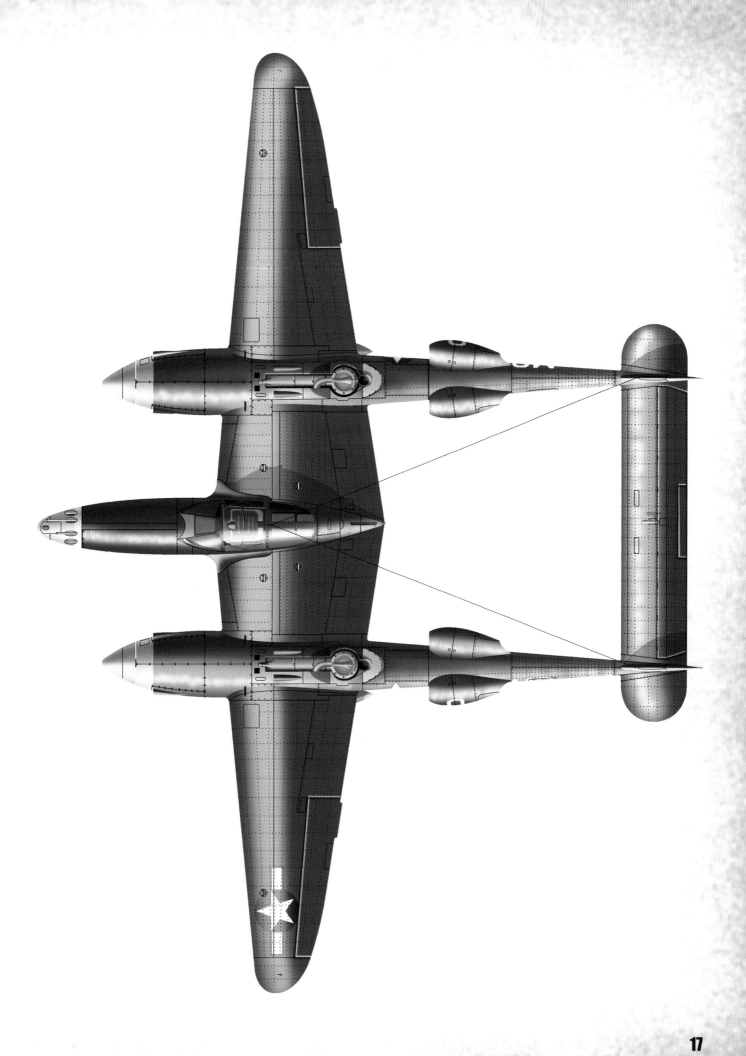

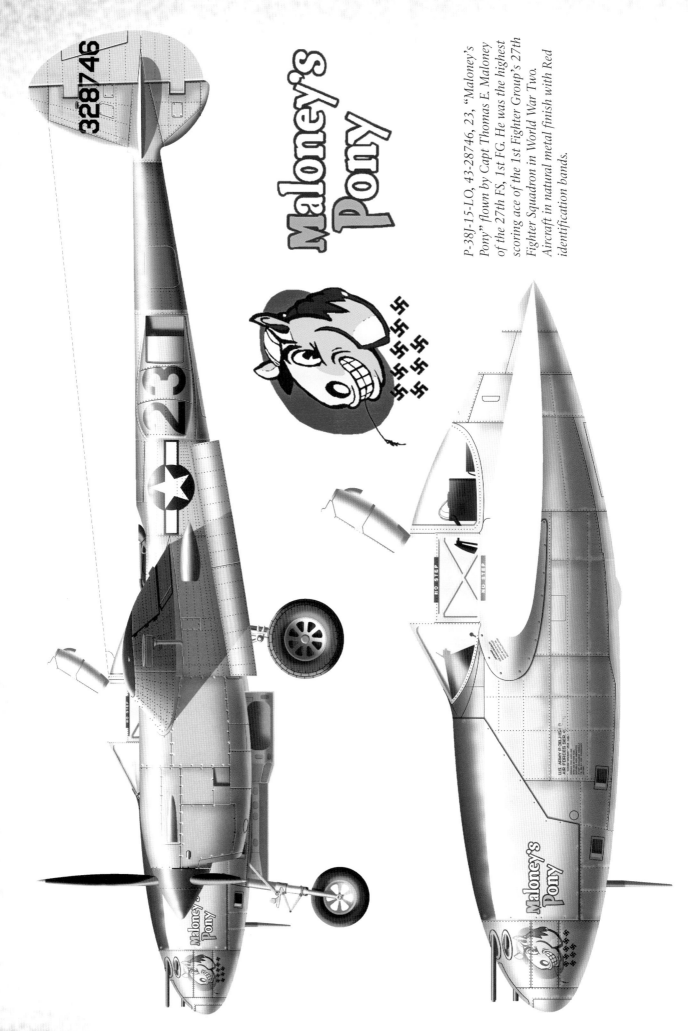

Maloney's Pony

P-38J-15-LO, 43-28746, 23, "Maloney's Pony" flown by Capt Thomas E. Maloney of the 27th FS, 1st FG. He was the highest scoring ace of the 1st Fighter Group's 27th Fighter Squadron in World War Two. Aircraft in natural metal finish with Red identification bands.

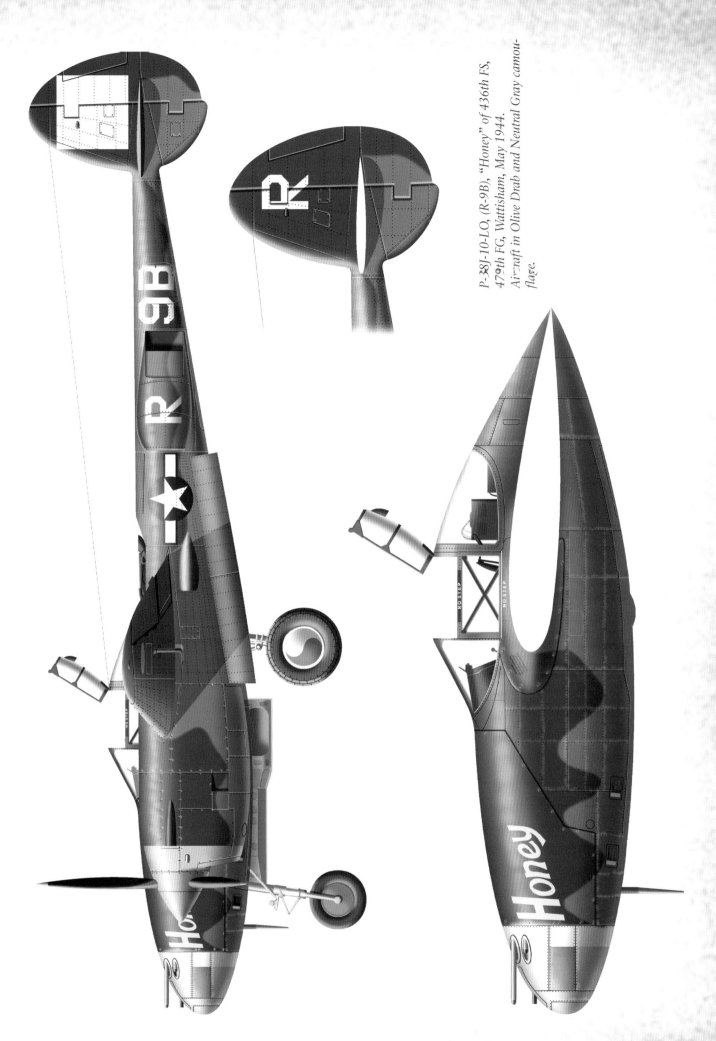

P-38J-10-LO, (R-9B), "Honey" of 436th FS, 479th FG, Wattisham, May 1944. Aircraft in Olive Drab and Neutral Gray camouflage.

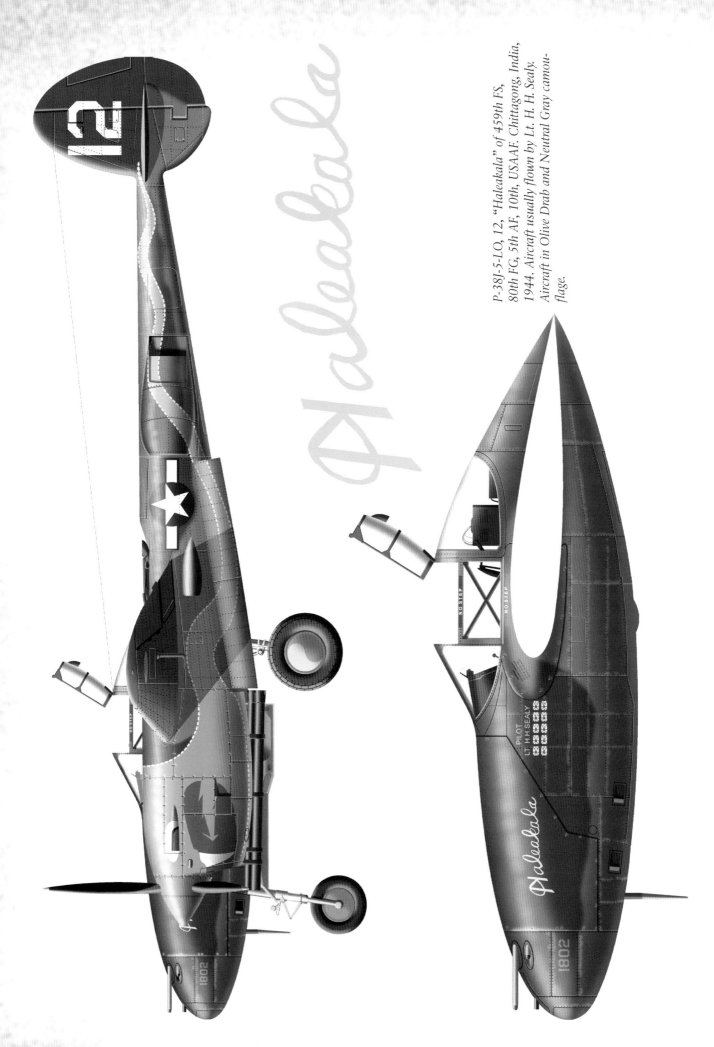

P-38J-5-LO, 12, "Haleakala" of 459th FS, 80th FG, 5th AF, 10th, USAAF. Chittagong, India, 1944. Aircraft usually flown by Lt. H. H. Sealy. Aircraft in Olive Drab and Neutral Gray camouflage.

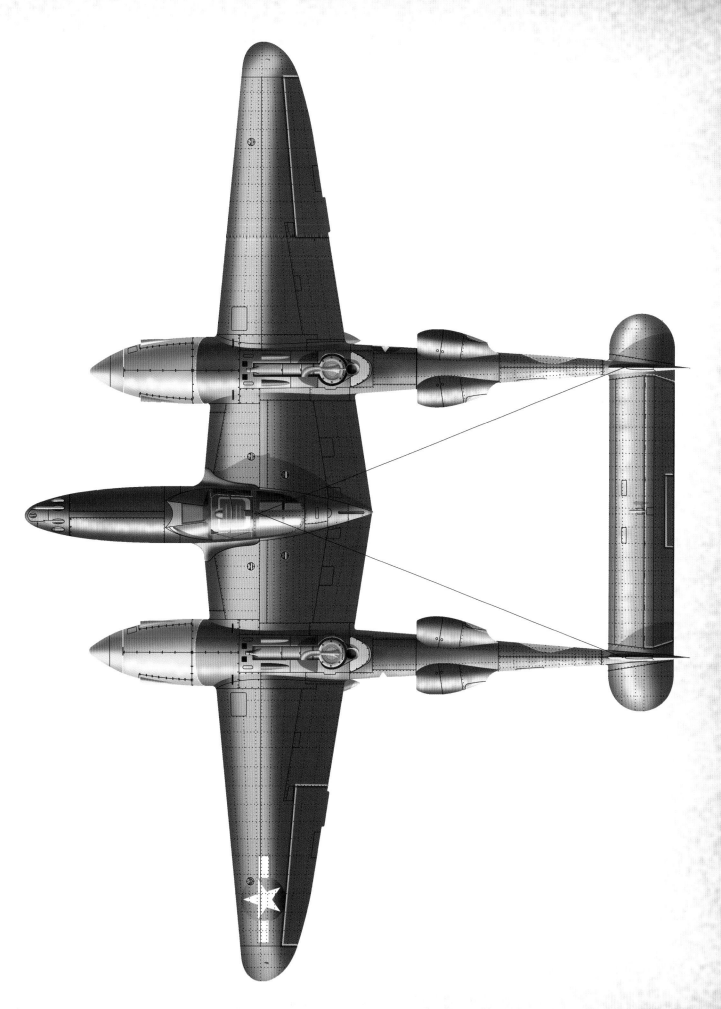

P-38J-15-LO, 43-28677, 677, B-HS, "Little Buckaroo" of 392nd FS, 367th FG. Flown by Maj. Robert C. "Buck" Rogers. Aircraft in natural metal finish with Red identification bands.

Maj. Robert C. "Buck" Rogers and his P-38J-25-LO Lightning. Photo was taken on 12 October, 1944 at Clastres, France (A-71). (US National Archives)

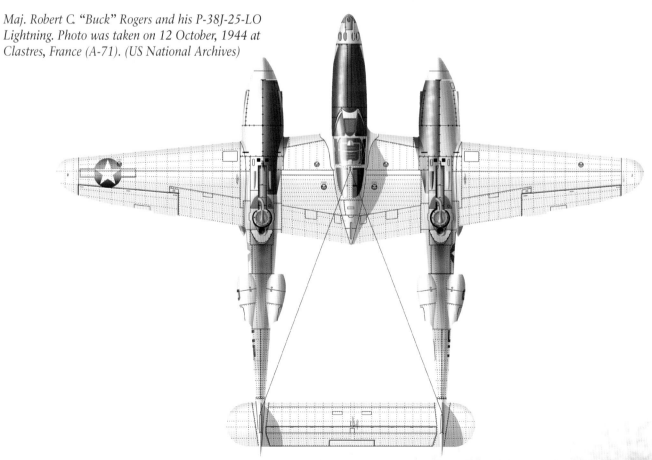

P-38K

This version was powered by 1,445 hp Allison V-1710-75/77 engines, driving Hamilton Standard propellers with slightly broader blades and a larger diameter (3.81 m). The aircraft offered only slightly better performance than the P-39J and, because the engine deliveries were not certain, no production was undertaken. Only one prototype was built.

P-38K-1-LO serial number 42-13558

P-38L

This version retained all the changes introduced in the P-38J. The new Allison V-1710-111/113 engine was rated at 1,495 hp for take off and 1,622 hp maximum combat (boosted) output. An improved turbosupercharger was fitted, and the fuel system and fuel pump were redesigned. For the first time additional tanks with pressure fuel feed were used, resulting in four bulged fairings under wings that housed the booster fuel pumps.

External changes included repositioning of the landing light to the leading edge of the port wing and of the gun camera to the port fixed bomb rack. This version was also able to be armed with rockets. Initially the launchers were attached directly under the wings, but firing the rockets caused wing skin deformation. Therefore, a special "Christmas Tree" launcher was developed.

The cockpit arrangement was changed (in fact it changed with virtually every version) and so was the radio equipment setup. The L version used the SCR-522 radio set, and a radio direction finder was also fitted.

The P-38L was built in two blocks:

P-38L-1-LO – 1,290 aircraft – virtually the same as the P-38J-25-LO, except for the new engines;

P-38L-5-LO – 2,520 aircraft – with the redesigned fuel system and rocket launcher. Also the attachments for the wing-mounted carriers were modified to be able to carry a total of two 2,000 lb. bombs or two 300 US-gallon tanks. AN/APS-13 warning radar was fitted in the lower section of the starboard fin;

P-38L-5-VN – 113 aircraft built under licence by the Consolidated-Vultee Aircraft Corporation plant at Nashville.

Serial numbers:	
44-23769—25058	Lockheed P-38L-1-LO
44-23059—27258	Lockheed P-38L-5-LO
44-53008—53327	Lockheed P-38L-5-LO
44-53328—54707	Lockheed P-38L-5-LO (contract cancelled)
43-50226—50338	Convair P-38L-5-VN
43-50339—52225	Convair P-38L-5-VN (contract cancelled)

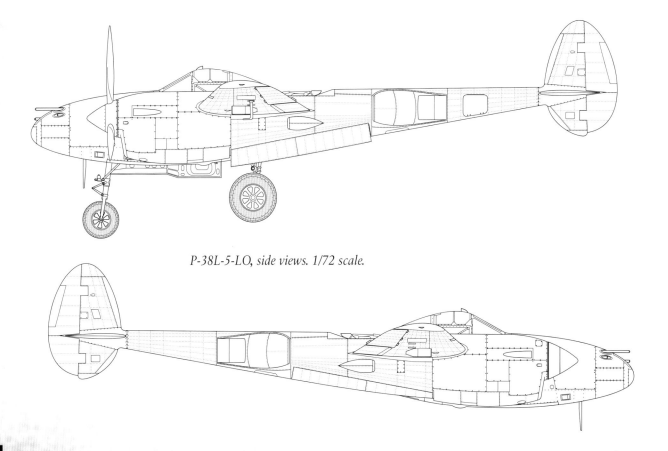

P-38L-5-LO, side views. 1/72 scale.

P-38L-1-LO, s/n 44-24008, "Little Red Head II" of 318th FG. (US National Archives)

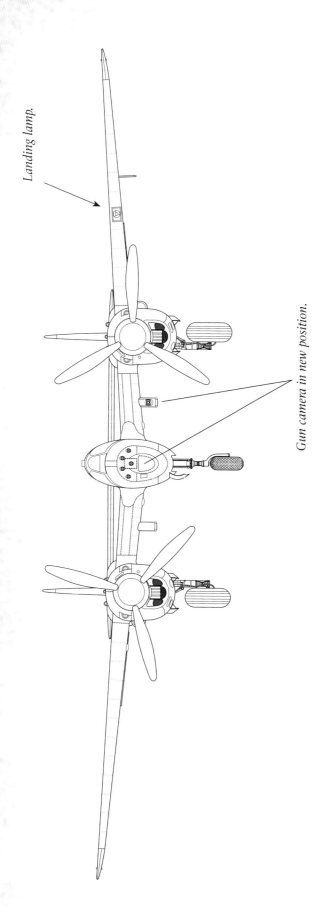

Landing lamp.

Gun camera in new position.

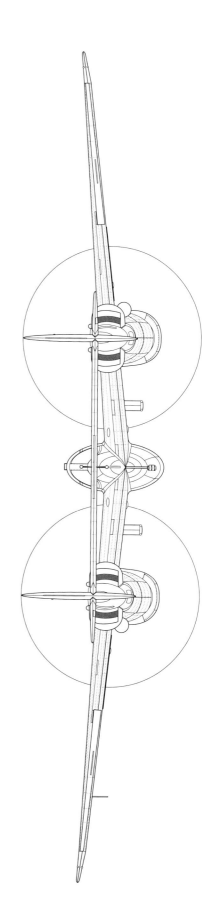

P-38L-5-LO front and rear elevation.

From nearest to farthest:
P-38L-1-LO s/n 44-24217;
P-38L-1-LO s/n 44-24379 Shot down by AAA over Padova,
Italy on 23 April,1945. Pilot was Capt. John D. Hurst;
P-38J-15-LO s/n 42-104428;
P-38J-15-LO s/n 43-28650.
Note that compressibility flaps are clearly visible.
(US National Archives)

Left: Scrap view of
AN/APS-13 warning
radar fitted in the
lower section of the
starboard fin.

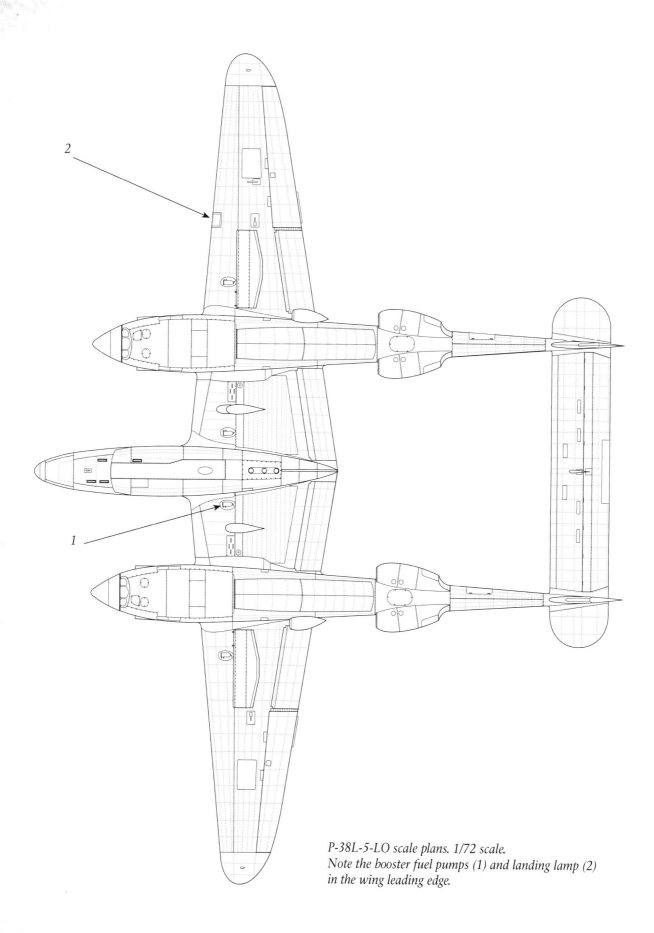

P-38L-5-LO scale plans. 1/72 scale.
Note the booster fuel pumps (1) and landing lamp (2)
in the wing leading edge.

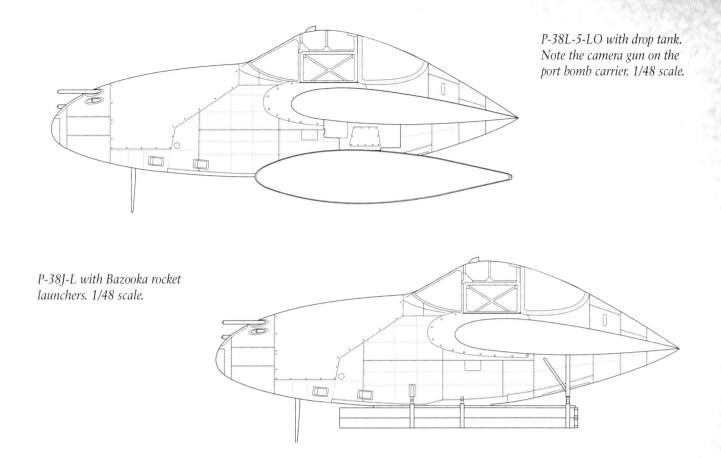

P-38L-5-LO with drop tank. Note the camera gun on the port bomb carrier. 1/48 scale.

P-38J-L with Bazooka rocket launchers. 1/48 scale.

P-38L "Hammer's Destruction Co." assigned to 1st Lt. Samuel E. Hammer of 90th FS, 80th FG, 10 AF. He has five confirmed aerial victories. India 1945. Note DF loop antenna under the nose. (US National Archives)

P-38L-5-LO with "Christmas Tree" rocket launchers.
1/72 scale.

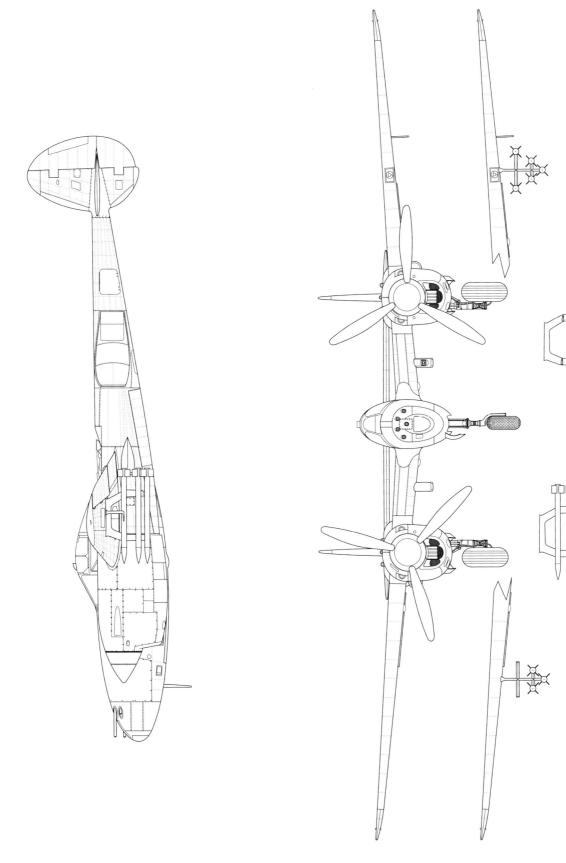

Above: Starboard side of the P-38L nose. P-38Js of 94th FS, 1st FG, Corsica 1944. *(US National Archives)*

Below: P-38L-5-LO, s/n 44-25734. "Betts II" of the 71st FS, 1st FG, Lesina, Italy. The ship was lost 15 April 1945 with Maj. Joseph Elliott. *(US National Archives)*

P-38J-5-LO, 135, "Pappy's Bier-die" of 431st FS, 475th FG. Philippines. Flown by Maj. R.I. Cline, Co of 431st FS. Aircraft in natural metal finish with red identification bands.

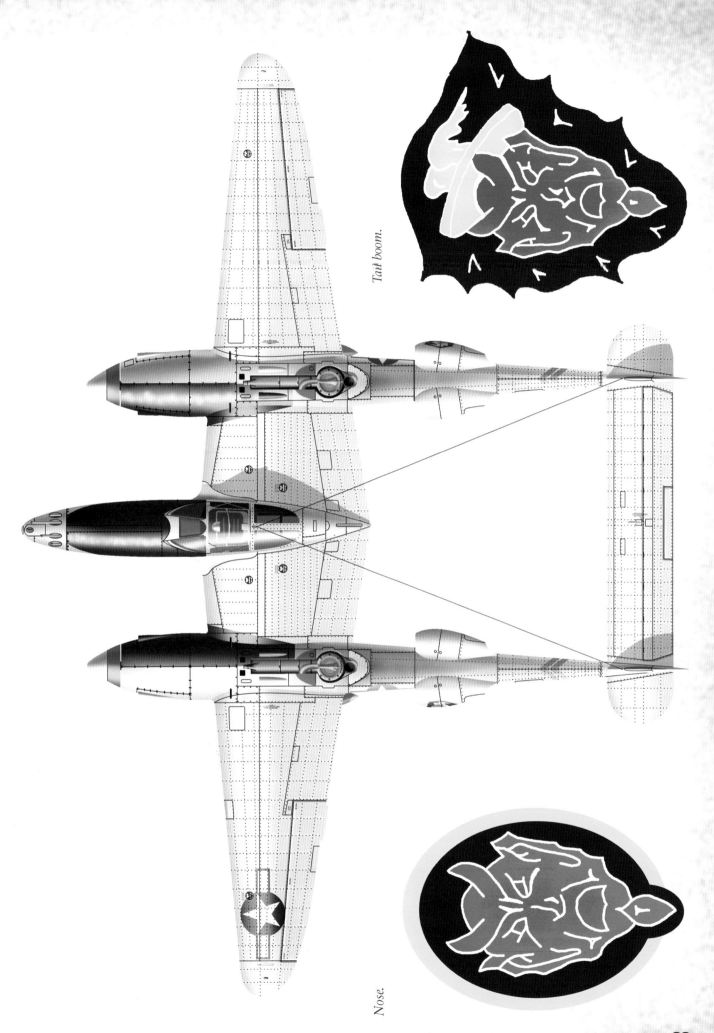

Tail boom.

Nose.

33

ALMOST "A" DRAGGIN

P-38J-5-LO, s/n 44-25638, "Almost A Draggin" of 9th FS, 49 FG, Lingayen, Philippines 1945. Aircraft was flown by Maj. Clayton M. Isaacson, who had scored four victories when he posed with his aircraft. Aircraft in natural metal finish.

"kittie"

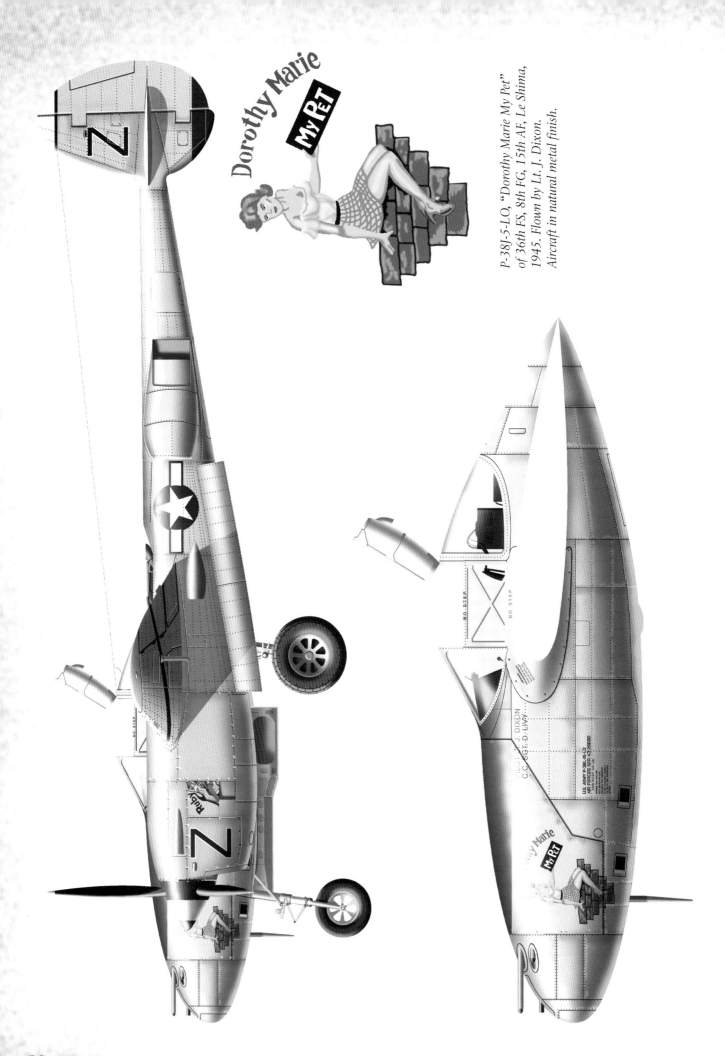

Dorothy Marie
MY PET

P-38J-5-LO, "Dorothy Marie My Pet"
of 36th FS, 8th FG, 15th AF, Le Shima,
1945. Flown by Lt. J. Dixon.
Aircraft in natural metal finish.

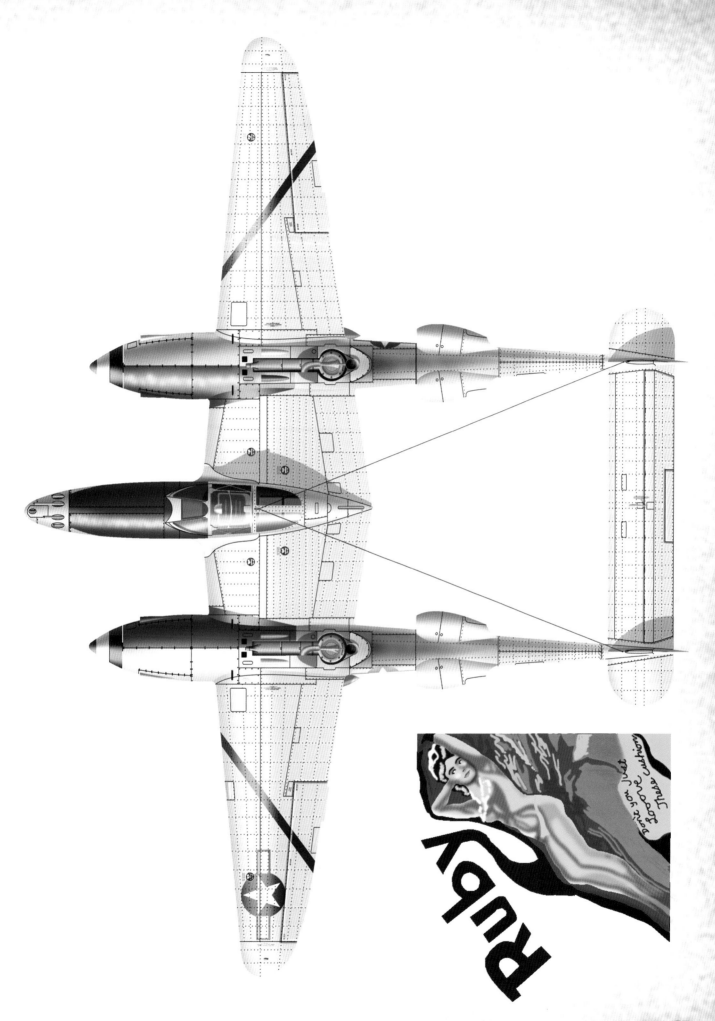

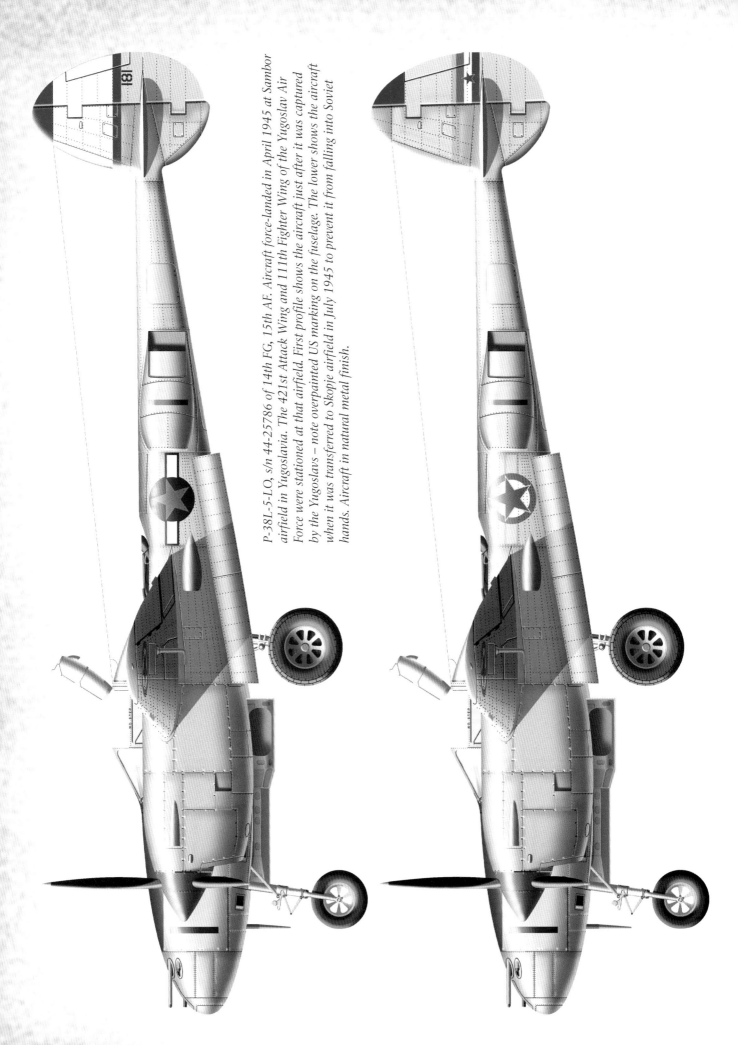

P-38L-5-LO, s/n 44-25786 of 14th FG, 15th AF. Aircraft force-landed in April 1945 at Sambor airfield in Yugoslavia. The 421st Attack Wing and 111th Fighter Wing of the Yugoslav Air Force were stationed at that airfield. First profile shows the aircraft just after it was captured by the Yugoslavs – note overpainted US marking on the fuselage. The lower shows the aircraft when it was transferred to Skopje airfield in July 1945 to prevent it from falling into Soviet hands. Aircraft in natural metal finish.

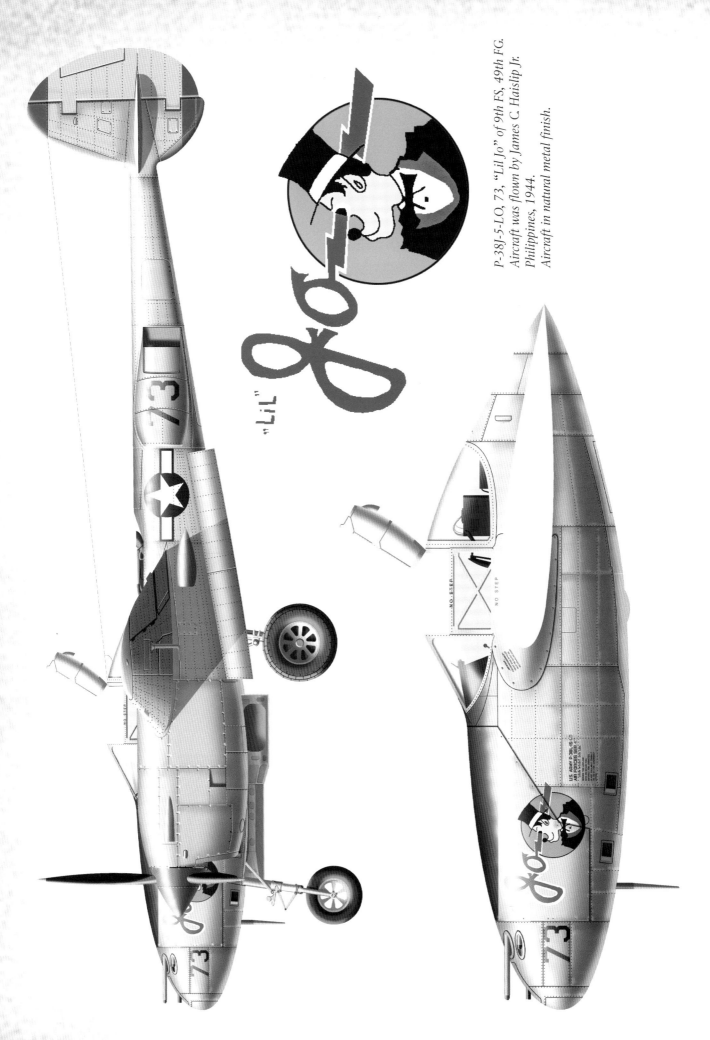

"LiL" Jo

P-38J-5-LO, 73, "Lil Jo" of 9th FS, 49th FG.
Aircraft was flown by James C. Haislip Jr.
Philippines, 1944.
Aircraft in natural metal finish.

PUTT PUTT MARU

P-38L-5-LO, s/n 44-25643, 100, "Putt Putt Maru", personal aircraft of Col. Charles McDonald, CO of 475th FG, Philippines 1945. Aircraft in natural metal finish.

P-38M

The excellent flying characteristics of the Lightning, and delays in the introduction of the P-61 Black Widow, led to an attempt to adapt the P-38 for night fighter duties. After a series of trials and experiments during late 1944 the USAAF placed a contract with Lockheed for conversion of a P-38L-5-LO, serial number 44-25237, to a two-seat night fighter. Conversion was made at Lockheed's Dallas Modification Center. After acceptance, the USAAF ordered 75 aircraft converted from the P-38L-5 version. The modified machines received the designation P-38M-5-LO.

The conversion consisted of fitting a radar operator 's compartment aft of the cockpit, and the radar in a streamlined pod under the nose of the aircraft. Armament was provided with special muzzle fairings to reduce glare from the gunflashes.

The aircraft were not used in combat, as they were too late in entering service with operational units.

Serial numbers of the aircraft converted to P-38M-5-LO.
44-26831, 26863, 26865, 26892, 26951, 26997, 26999, 27000, 27108, 27233, 27234, 27236, 27237, 27238, 27245, 27249, 27250, 27251, 27252, 27254, 27256, 27257, 27258, 53011, 53012, 53013, 53014, 53015, 53016, 53017, 53019, 53020, 53022, 53023, 53025, 53029, 53030, 53031, 53032, 53034, 53035, 53042, 53050, 53052, 53056, 53062, 53063, 53066, 53067, 53068, 53069, 53073, 53074, 53076, 53077, 53079, 53080, 53082, 53083, 53084, 53085, 53086, 53087, 53088, 53089, 53090, 53092, 53093, 53094, 53095, 53096, 53097, 53098, 53100, 53101, 53106, 53107, 53109, 53110, 53112.

Two photos of the P-38M-LO with AN/APS-4AI radar pod. Machine gun and cannon muzzles have flame dampers installed. Aircraft of 319 NFS, Hammer Field, summer 1945. (US National Archives)

Radar operator's cockpit showing radar screen. (US National Archives)

P-38M main fuselage, 1/48 scale.

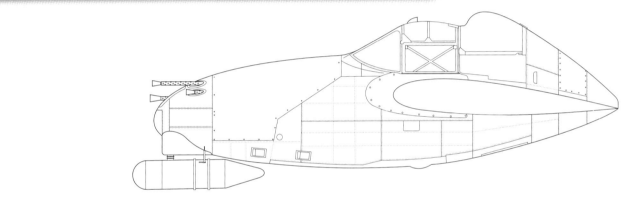

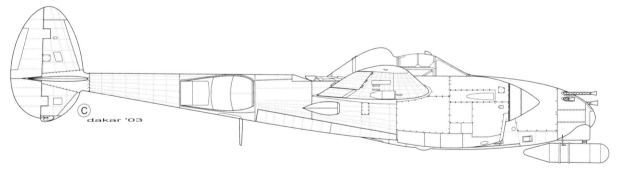

dakar '03

P-38M side views, 1/72 scale.

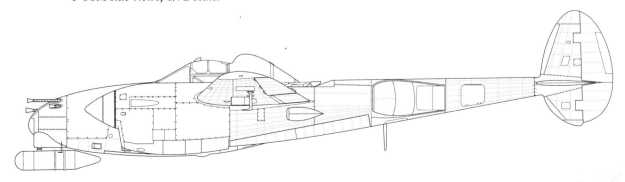

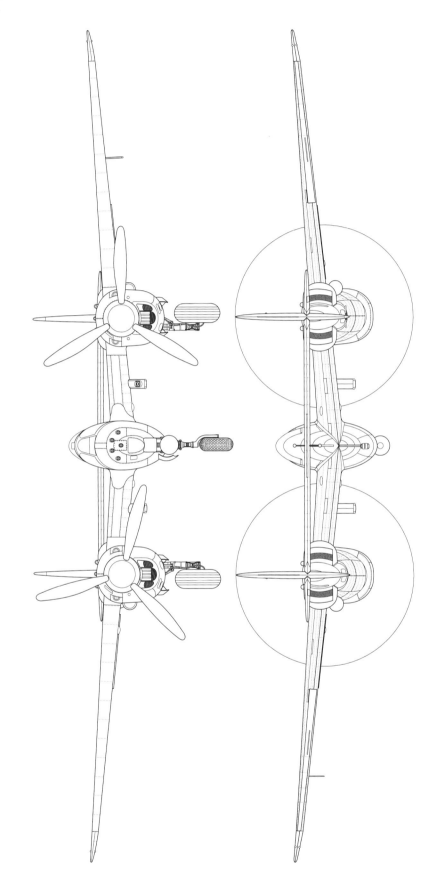

P-38M front and rear views, 1/72 scale.

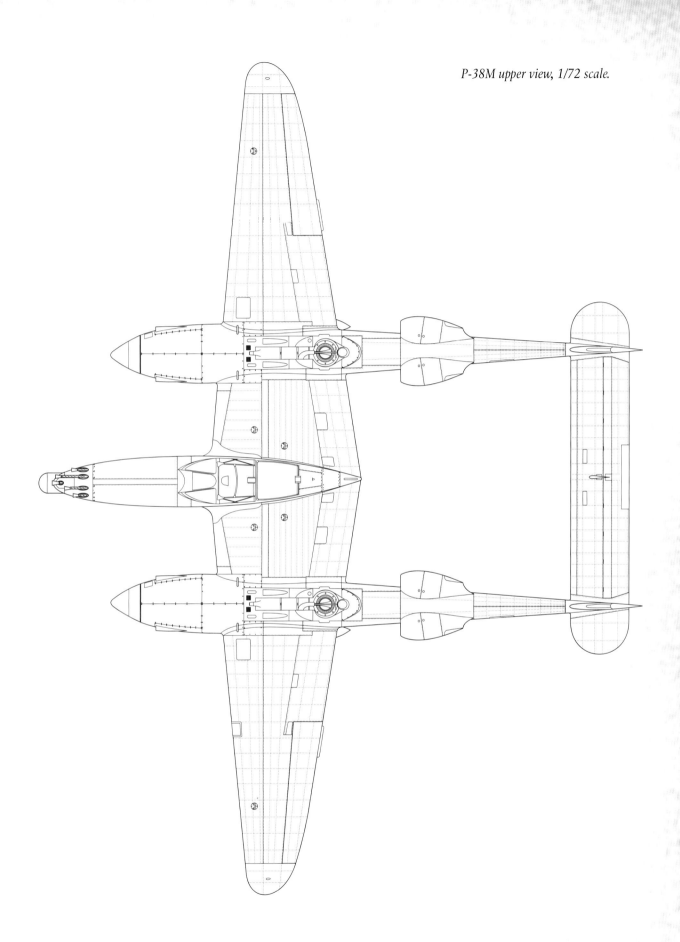

P-38M upper view, 1/72 scale.

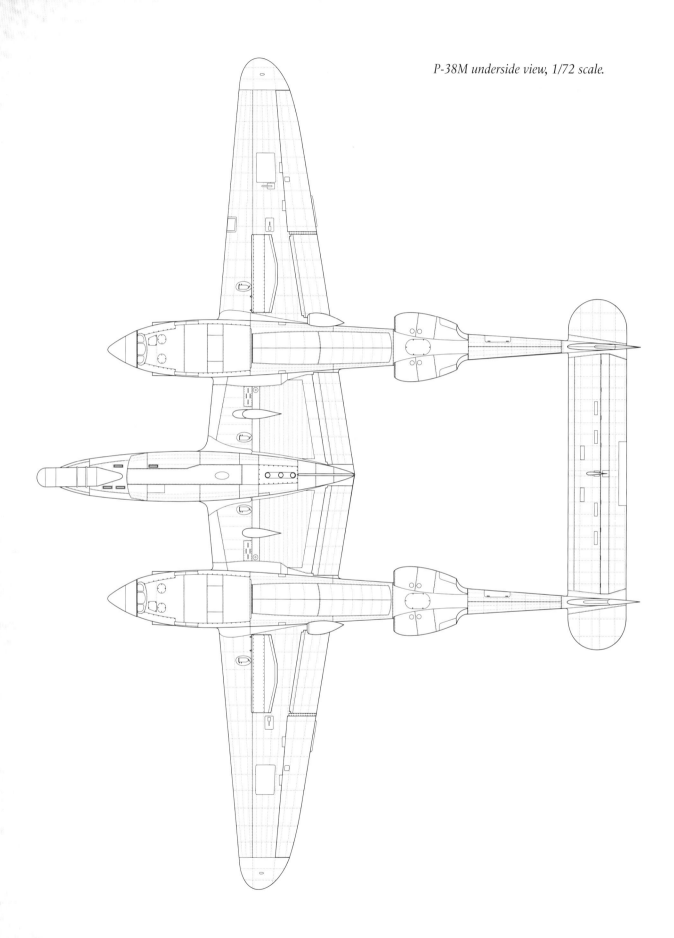

P-38M underside view, 1/72 scale.

P-38M-5-LO, s/n 44-27245, "Shady Lady/Snooks", Atsugi, Japan 1945. Pilot 1/Lt. Johny Brewer, radar operator 1/Lt. David Hopwood.

P-38J "Droop Snoot" (model 422-81-14)

As the bomb load and range of late P-38 versions was equal to that of the B-17, the idea to use the P-38 as a strategic bomber was put forward for obvious reasons (faster aircraft, fewer air crew, lower fuel consumption). Such aircraft needed a bomb-sight or had to be guided by an aircraft with a bomb sight. This version of the P-38 was obtained by fitting an extensively glazed bomb aimer's compartment in the fuselage nose. The station was fitted with a Norden sight. Trials were conducted using a P-38H, serial number 42-67086. The converted aircraft would lead Lightnings each loaded with 4,000 lb. of bombs, dropping on the "Droop Snoot" bombardier's command.

After successful tests Lockheed converted 23 P-38Js. The aircraft were known colloquially as the P-38J "Droop Snoot". Moreover, 100 conversion kits were manufactured and despatched to units.

The first bombing raid with use of a P-38J "Droop Snoot" was carried out on 10 April 1944, and the type was used exclusively in Europe.

A "Droop Snoot" (P-38J) of unknown unit, France 1944.
(Renita)

A "Droop Snoot" (P-38J) of the 402nd FS, 370th FG, 9th AF taking off from a base on 10 April 1945. (US National Archives)

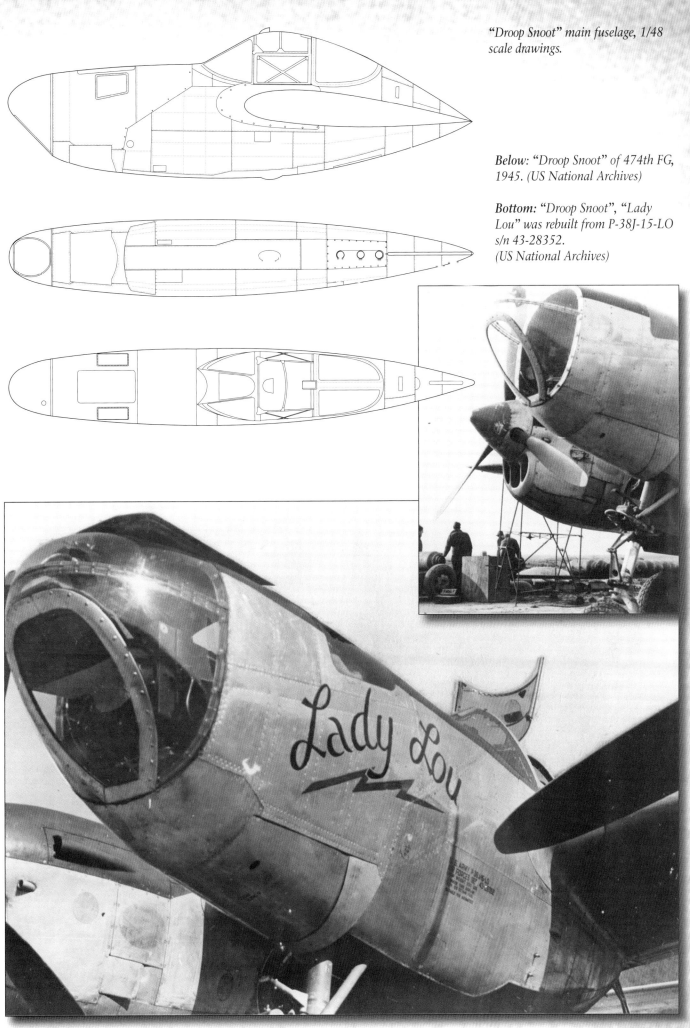

"Droop Snoot" main fuselage, 1/48 scale drawings.

Below: "Droop Snoot" of 474th FG, 1945. (US National Archives)

Bottom: "Droop Snoot", "Lady Lou" was rebuilt from P-38J-15-LO s/n 43-28352. (US National Archives)

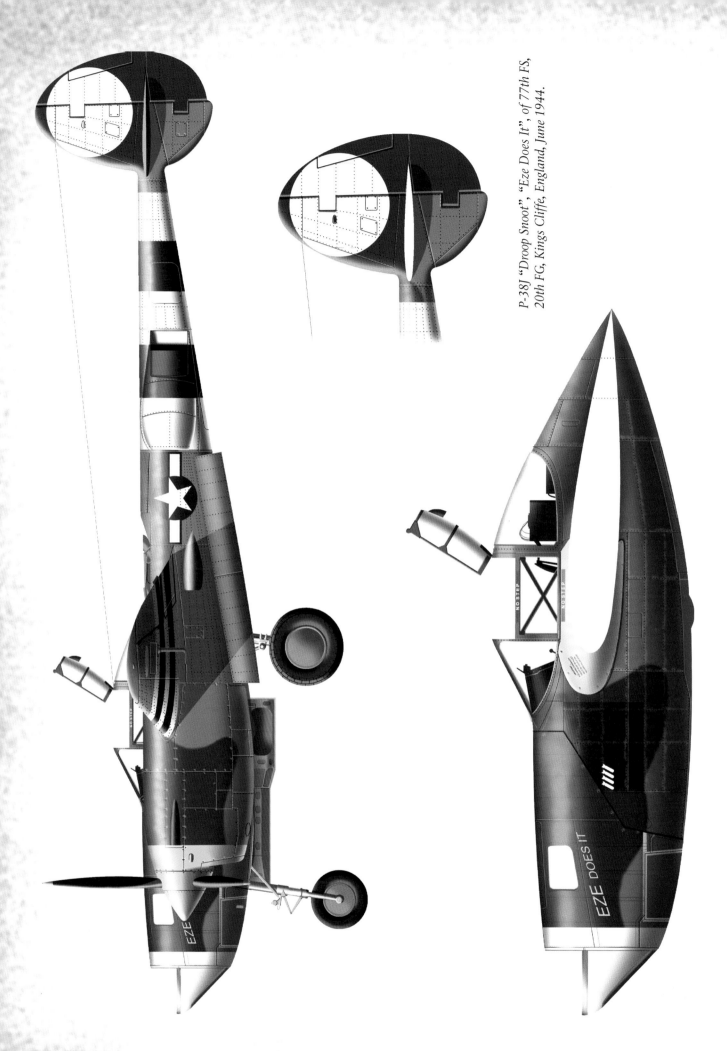

P-38J "Droop Snoot", "Eze Does It", of 77th FS, 20th FG, Kings Cliffe, England, June 1944.

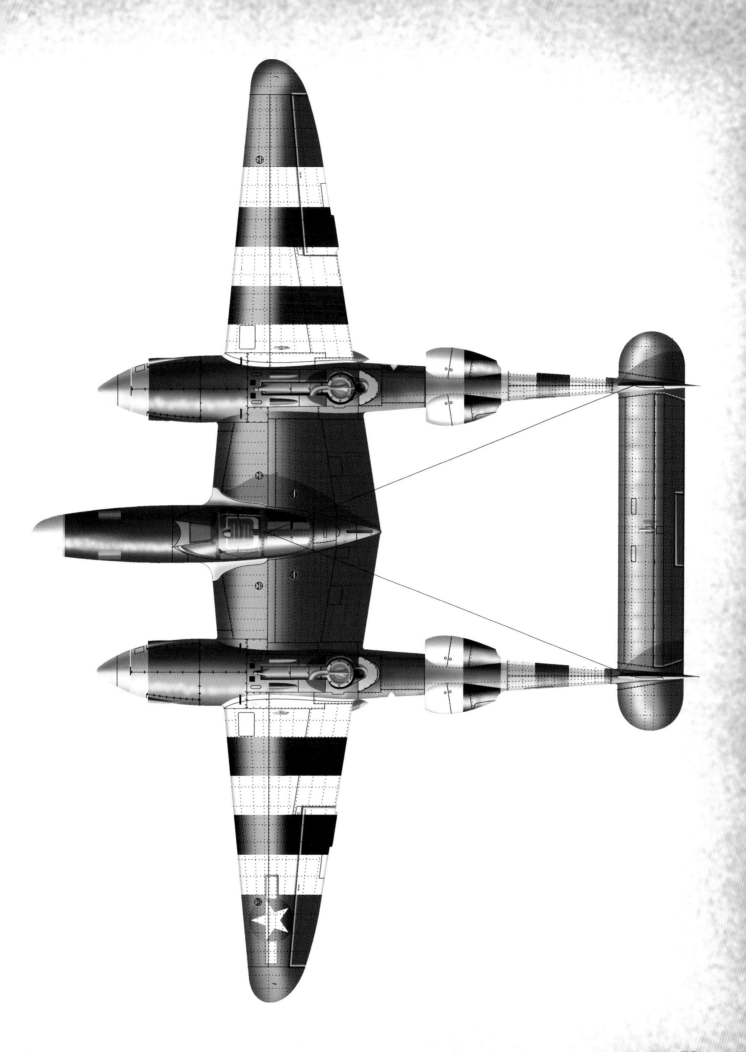

P-38 "Pathfinder" (Model 422-81-22)

This version was virtually identical to the "Droop Snoot", but converted from the P-38L, and fitted with AN/APS-15 radar for detecting ground targets in bad weather.

Above: P-38L "Pathfinder" converted from P-38L-1.
Below: "Pathfinder" nose with radar operator doors in open position. Both (US National Archives)

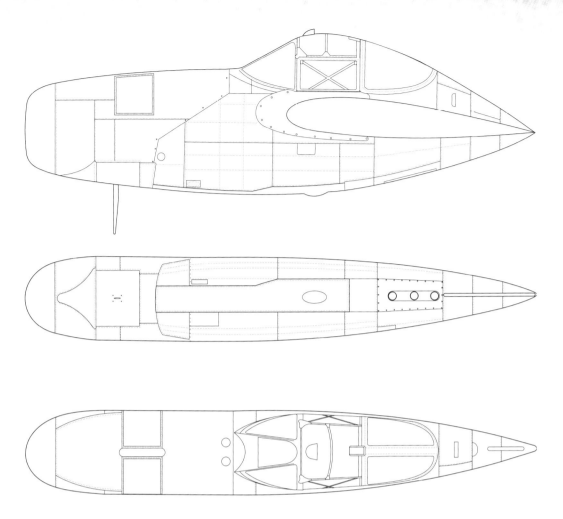

P-38L "Pathfinder" – fuselage. 1/48 scale plans.

P-38J-15-LO, s/n 44-23139 modified at factory to "Pathfinder" version with AN/APS-15 radar. (US National Archives)

Reconnaissance versions

Earlier versions of the P-38 had already been converted for reconnaissance duties by fitting cameras in place of the armament. The J – L versions described in this book were also converted into reconnaissance P-38 variants, as follows:

F-5B-1LO (Model 422-81-21)

This version was based on P-38H-1 and P-38J-10 aircraft. Various camera sets were used. The most frequent one consisted of two 6 in. focal length K-17 cameras for oblique photos and a 20 in. or 24 in. K-15 for vertical photos. These aircraft were fitted with a Sperry automatic pilot.

The F-5B-1-LO was the last version which was built as reconnaissance machine from the outset, later ones being converted from fighter P-38s after these were assembled and delivered to the USAAF. Two hundred aircraft were built, of which 110 were based on the P-38J-10-LO.

Serial numbers:
42-68192/68301- F5-B-1-LO
(P-38J-10-LO based).

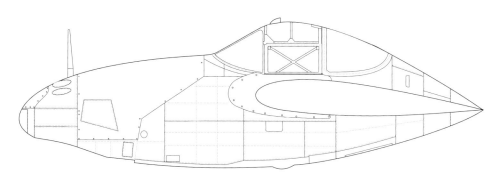

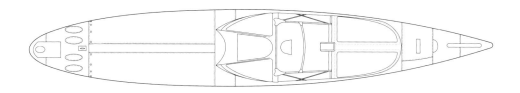

1/48 scale plans.

F-5B 42-68226 "Junior" of the 28th PRS taxies out for takeoff from Yontan airstrip, Okinawa, 10 July 1945. (US National Archives)

F-5B-1LO, s/n 42-68216 "Little Buff" of 31st PRS 10th PRG. Flown by F/O Iris L. Dillon. (US National Archives)

F-5C-1-LO (P-38J-5LO)

This designation applied to P-38J-5LO aircraft converted to reconnaissance aircraft. They were converted at Lockheed's Dallas Modification Center to a reconnaissance standard similar to the F-5B-1-LO, but with an improved photo system. A total of 128, or according to other sources 123, aircraft were converted. Because these machines were not listed by the manufacturers as reconnaissance machines, serial numbers were not recorded at assembly.

F-5E-2-LO (P-38J-15-LO)

Two hundred P-38J-15-LO aircraft were converted at Lockheed's Dallas Modification Center to a reconnaissance standard similar to the F-5C-1-LO. Because these machines were not listed by the manufacturers as reconnaissance machines, serial numbers were not recorded at assembly.

F-5E-3-LO (P-38J-25-LO)

Similarly, 105 P-38J-25-LO aircraft were converted to a reconnaissance version virtually identical to the F-5C-1-LO.

Serial numbers are unknown.

Above: F-5C-1-LO s/n 42-67128, "Dot And Dash" after landing at Eurk [Poltava?], Russia. (US National Archives)

Installing cameras in an F-5C fuselage.
(US National Archives)

F-5E-4 (P-38L-1-LO)

This version was converted from the P-38L-1-LO. It was equipped with four cameras. A total of 508 aircraft were converted.

F-5E s/n 44-2327 (?), "Ruth", clearly showing the camera windows. Note also mission markings. (US National Archives)

Another F-5E with similar mission markings. (US National Archives)

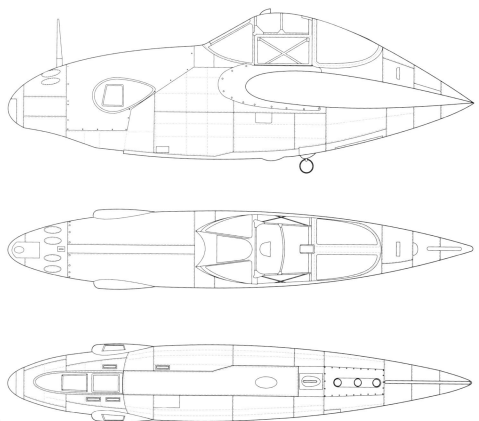

1/48 scale plans.

F-5F-3-LO (P-38L-5-LO).

These were P-38L-5-LO conversions with five cameras. The number of those modified is not accurately known.

Rare photo of the F-5F-3-LO, s/n unknown. (US National Archives)

1/48 scale plans.

Camera switch box mounted in F-5 cockpit. (US National Archives)

F-5G-6-LO (P-38L-5-LO)

This was the last reconnaissance version of the Lightning. It featured an enlarged nose compartment with four vertical cameras plus a single oblique camera at the front.

63 aircraft were converted at Lockheed's Dallas Modification Center.

Because these machines were not listed by the manufacturers as reconnaissance machines, serial numbers were not recorded at assembly.

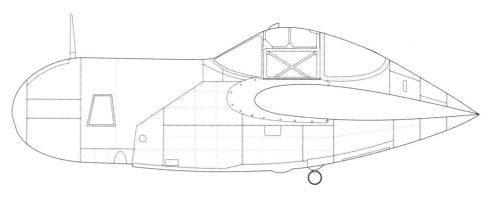

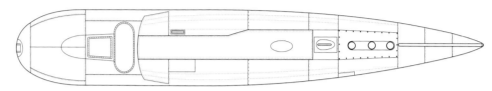

1/48 scale plans.

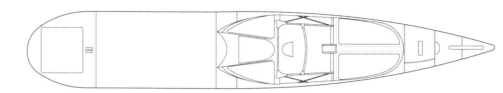

Free French F-5G-6-LO, L, 121. Note unit emblem on the nose. (via J. Fernandez)

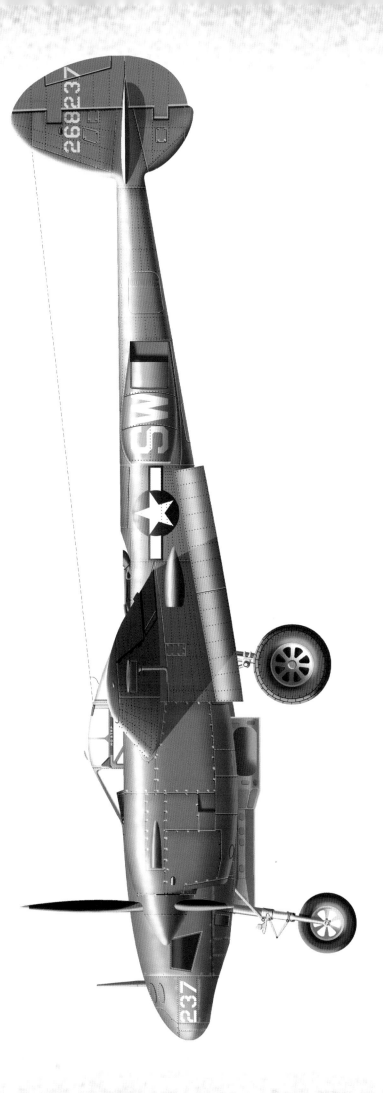

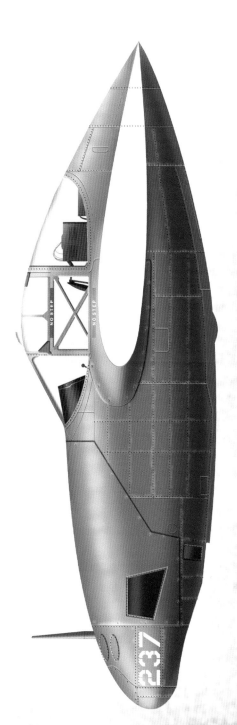

F-5B-LO, s/n, 42-68237, (SW), 237 of 33rd RS,
10RG Chalgrove, England, May 1944.
Synthetic Haze Paint camouflage.

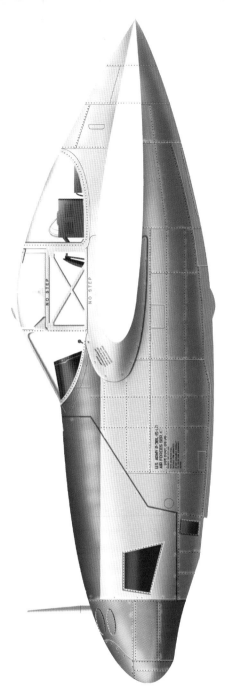

F-5B-LO, s/n 42-68207, Z-XX of 34th PRS
69th TRG, Haguenau, April 1945.
Aircraft in natural metal finish.

*F-5E-LO, s/n 44-25058, "8" of 8th PRS, 6th PRG,
Dulag Island, Philippines, 1945.
Aircraft in natural metal finish.*

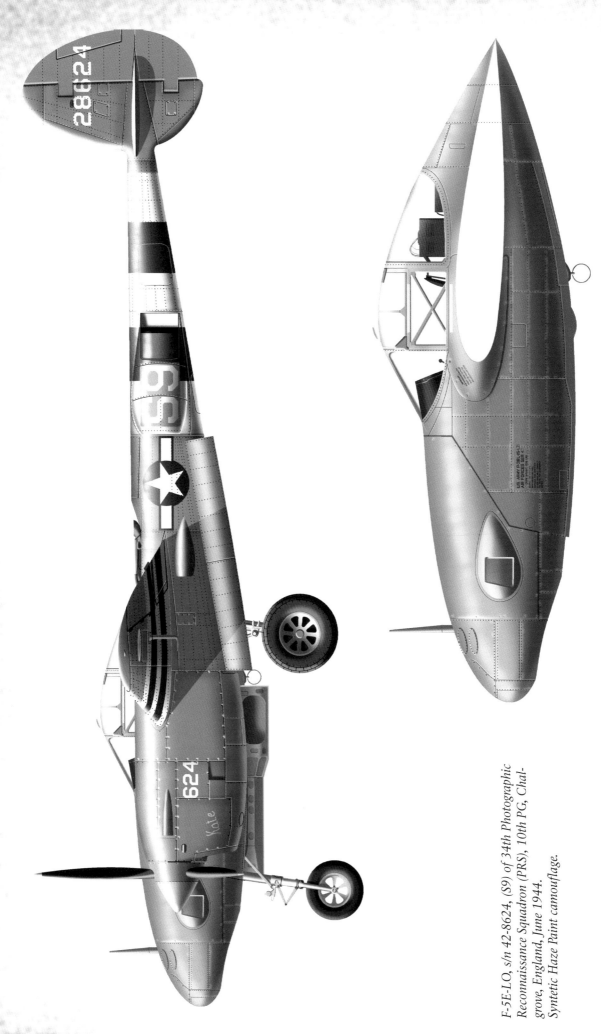

F-5E-LO, s/n 42-8624, (S9) of 34th Photographic Reconnaissance Squadron (PRS), 10th PG, Chalgrove, England, June 1944. Syntetic Haze Paint camouflage.

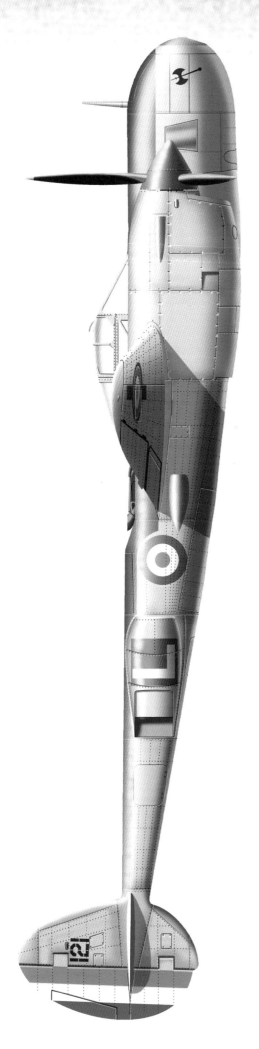

Above: Free French F-5G-6LO, s/n unknown, -L of GR 2/3, 1945. Aircraft in natural metal finish. French roundels in six positions.

Below: Free French F-5G-6LO, s/n 44-25915, (W4-Y) of GR 2/3, Fribourg, 1945. Aircraft in natural metal finish. French roundels in six positions.

Technical description

Single-seat, twin-engined, cantilever mid-wing monoplane of all-metal construction with retractable undercarriage, designed in twin-boom layout with a central fuselage pod that housed the cockpit and armament.

The central fuselage (pilot's pod) of semi-monocoque design with oval cross-section, with smooth duralumin covering. The skin thickness was between 0.025" (0.64 mm) and 0.051" (1.3 mm). The nose section housed the built-in armament and ammunition boxes. Access to it was provided by upward-opening covers on both sides of the fuselage. A VHF antenna was attached at the bottom nose covering. The pilot's cockpit was located aft of the frame that separated the armament compartment. The heavily glazed cockpit canopy (its profile merged in with the wing trailing edge) provided rearward visibility.

The cockpit canopy was of three-section design. The front and rear sections were fixed, while the top of the central section opened upwards and backwards, hinged on the duralumin reinforcement aft of the pilot's seat, and it could be jettisoned in emergency.

Side panels could be lowered using hand cranks. The windscreen included a bullet-proof glass panel at the front. (In the early versions the bullet-proof glass was an integral part of the windscreen, and constituted its front pane.) It was bonded with vinyl adhesive from five layers of glass. Side panels of the canopy were made from twin layers of glass bonded with the same adhesive. A ladder was fitted for entry into the cockpit and it retracted into the lower part of the pod. The nose wheel leg was fitted in the forward nose, attached on a special joint to the diagonal reinforced frame that separated the armament compartment from the cockpit. When retracted the nose wheel was located under the cockpit floor.

The cockpit was armour-protected and was fitted with a full set of flying and navigation instruments that allowed flying in adverse weather. The engine, propeller and undercarriage panel was located on the port side. The flap lever and radio panels were located on the starboard side.

The radio blocks were located on a shelf aft of the pilot's seat, immediately below the transparent hood. The rudder pedal bar was fitted in the floor, below the instrument panel, with adjustment of distance from the pilot's seat. Undercarriage brake levers were mounted in the rudder pedals. The yoke was attached to a control column that was pivoted under the floor on the starboard side. The pilot's seat, pressed from a laminate, included a recess for the parachute. Its back section included a 0.375" (ca. 9.5 mm) armour plate of hardened steel. Pilot's head was protected by a plate of similar material.

Identification lights were fitted in the bottom of the fuselage on the centreline.

The three-piece wing, cantilever, two-spar, all-metal, was of tapered planform with rounded tips, covered with smooth duralumin covering, its thickness ranging between 0.020" (ca. 0.5 mm) and 0.040" (ca. 1 mm). The wing tips were highly flattened on the under side. The wing airfoil was NACA 23016 at the root (aircraft centreline) changing to NACA 4412 at wing tips. The wing centre section was an integral structure with the cockpit and the engine nacelles. Outer wing panels were attached to the centre section at the wing-engine nacelle joint. In late J and L version they were fitted with air brakes. These consisted of small rectangular hinged metal panels, on the underside of the wings. They served to reduce speed and allowed quick recovery from a dive.

Metal ailerons, covered with smooth duralumin, deflected differentially with hydraulic actuation. Four weights were attached to the aileron spar for mass balance. Similar design of the aileron attachment was rarely seen on other combat aircraft: the aileron was attached to the upper wing surface on a piano hinge along its entire span. Each aileron was fitted with a balance tab adjusted on the ground and a trim tab adjustable in flight (the latter were deleted on later variants).

Fowler flaps of metal construction covered the wing span from the pilot's pod to the ailerons, with a break for the engine booms.

Four integral self-sealing fuel tanks with a total capacity of 1,136 litres (in later version) were fitted inside the wing centre section. Tanks for 208 litres were fitted in later versions in the outer wings forward of the spar.

A night landing light was fitted in the leading edge of the port wing. Navigation lights were fitted on the upper and lower wing tip surfaces.

Attachments for external stores (bombs, torpedoes) or fuel tanks were located symmetrically under the wing centre section. Starting from the J version rocket missile carrier attachments were fitted under outer wings.

The Pitot tube was located under the port wing.

Tail: cantilever, of all-metal construction, with smooth duralumin covering.

Three-piece horizontal tail, with rectangular planform with semi-circular tips extending beyond the engine booms. Single-piece elevator, mass balanced, fitted with a balance tab adjustable in flight from the cockpit.

Twin vertical tail surfaces, with characteristic oval planform. The shape of the section above the tail boom was an ogival ellipse, while the section below the tail boom was semi-circular with a small steel skid protruding. The fin, split in the same way as the rudder, consisted of the upper section (above the boom) and the bottom section (below the boom). Aerodynamically balanced rudders were fitted with balance tabs adjustable in flight. White navigation lights were fitted on outboard sides of the fins.

Rudders, elevator, ailerons and trim tabs were controlled by steel cables.

Tricycle undercarriage, completely retractable and faired over in flight with covers that completed the outer streamlined form of the fuselage and booms. Main wheels were retractable backwards into the engine booms, and the nose wheel into the pilot's pod. The nose wheel was fitted with a vibration damper. The oleo legs were hydraulically retracted and lowered. The main wheels were fitted with drum brakes and a parking brake.

Power plant consisted of two Allison V-1710 liquid-cooled 12-cylinder Vee in-line engines, with cylinder rows banked at 60°, with various engine variants depending on the version of the aircraft.

Cylinder diameter was 5.5 inch (139.7 mm), piston stroke was 6 inch (152.4 mm). Cubic capacity was 1,710 cu. in. (28 litres), compression ratio 6.65:1.

Engine specification Allison V-1710-111/113 (F30R/L)	
Fuel	100/130-octane
Take-off power	1,475 hp at 3,000 rpm
Normal power @ altitude	1,100 hp at 2,600 rpm @ 30,000 ft (9.150 m)
Combat power (15-minute) @ altitude	1,475 hp at 3000 rpm @ 30,000 ft (9.150 m)
Empty engine weight	1,395 lbs (633.5 kg)
Dimensions: length, height, width	85.91" (2.180 m), 36.65" (0.958 m), 29.28" (0.744 m)

Armament of the aircraft consisted of a single 20 mm AN-M-Z cannon and four 0.5 in. (12.7 mm) M-2 machine guns. The machine gun positions were staggered to allow greater amount of ammunition. Bigger ammunition boxes could be located one behind another rather than in pairs side-by-side. Normal ammunition load was 300 rounds per gun, but a maximum of 500 rounds per gun could be loaded into each box.

Blast tubes were fitted over the protruding sections of machine gun barrels. Aircraft from the serial range of 42-66502 to 42-67311 had these tubes with 45 mm outer diameter. Aircraft from serial number 42-67402 onwards were fitted with tubes of a slightly smaller diameter.

The aircraft was able to carry a maximum load of 1,450 kg bombs on carriers under the wing centre section. There were other modifications that allowed the P-38 to carry three-tube Bazooka launchers.

Late versions of the P-38 were also fitted with attachments to carry 5 in. (127 mm) HVAR missiles. These modifications were initially made in workshops; later on, in the L version, they were factory-fitted.

P-38J-25-LO Lightning, s/n 44-23569,"Curly Six" of 394th FS, 367th FG, 9th AF. Aircraft assigned to Maj. Jack L. Reed. (US National Archives)

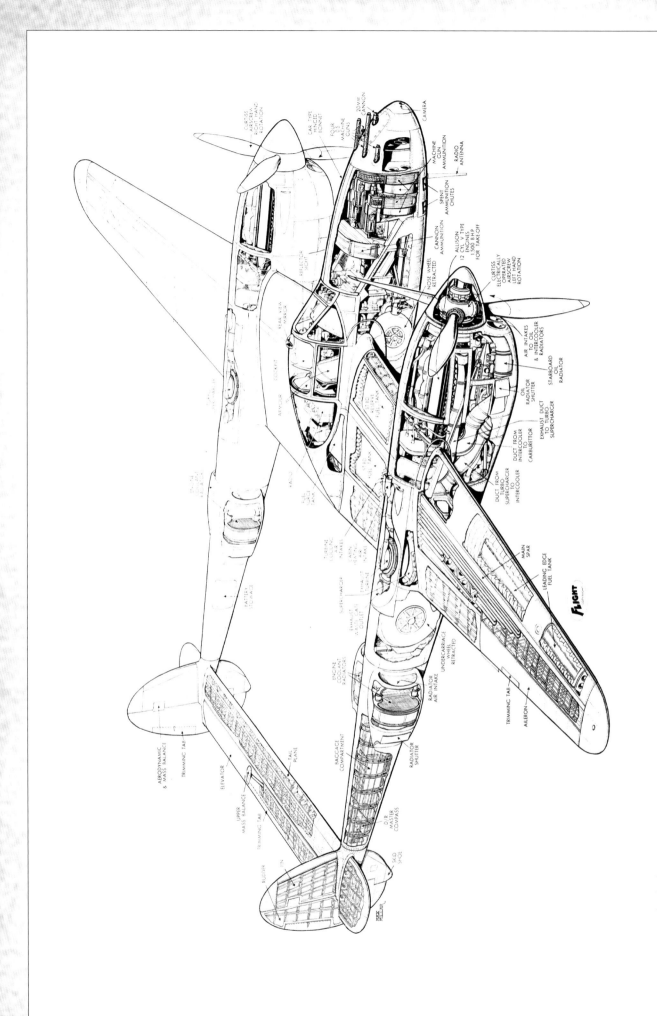

P-38J cutaway published in Flight *in 1945.*

General characteristics P-38

Dimensions
Length: 37 ft 10 in (11.53 m)
Wingspan: 52 ft 0 in (15.85 m)
Height: 12 ft 10 in (3.91 m)
Wing area: 327.5 ft² (30.43 m²)
Airfoil: NACA 23016 / NACA 4412
Empty weight: 12,800 lb (5,800 kg)
Loaded weight: 17,500 lb (7,940 kg)
Max. takeoff weight: 21,600 lb (9,798 kg)
Zero-lift drag coefficient: 0.0268
Drag area: 8.78 ft² (0.82 m²)
Aspect ratio: 8.26[118]

Performance
Maximum speed: 443 mph (713 km/h) (on War Emergency Power: 1,725 hp at 64 in HG and 28,000 ft (8,530 m)
Stall speed: 105 mph (169 km/h))
Range: 1,300 mi (2,100 km) combat (1,770 km / 3,640 km)
Service ceiling: 44,000 ft (13,400 m)
Rate of climb: 4,750 ft/min (24.1 m/s) maximum
Wing loading: 53.4 lb/ft² (260.9 kg/m²)
Power/mass: 0.16 hp/lb (0.27 kW/kg)
Lift-to-drag ratio: 13.5

Armament
1× Hispano M2(C) 20 mm cannon with 150 rounds
4× Browning MG53-2 0.50 in (12.7 mm) machine guns with 500 rpg.
4× M10 three-tube 4.5 in (112 mm) rocket launchers; or:
Inner hardpoints:
2× 2,000 lb (907 kg) bombs or drop tanks; or
2× 1,000 lb (454 kg) bombs or drop tanks, plus either
4× 500 lb (227 kg) bombs or
4× 250 lb (113 kg) bombs; or
6× 500 lb (227 kg) bombs; or
6× 250 lb (113 kg) bombs
Outer hardpoints:
10× 5 in (127 mm) HVARs (High Velocity Aircraft Rockets); or
2× 500 lb (227 kg) bombs; or
2× 250 lb (113 kg) bombs

P-38J-10-LO, s/n 42-68092, "In Memory of Lt. F. Slanger, U.S.A.N.C." of 392nd FS, 367th FG, 9th AF, flown by 2nd Lt. Aubrey J. Oldham. (US National Archives)

DETAIL PHOTOS
GENERAL VIEW

P-38J, s/n 42-67543 in flight at a UK airshow.. (J. Kightly)

P-38J-15-LO, s/n 43-28430, (KI-N), "Jeanne" of 55th FS, 20th FG. Capt T. Roy M. Scrutchfield with his ground crew. This aircraft was lost in a flying accident on 16 June 1944 with Maj. Paul A. Lobinger at the controls. (US National Archives)

Rear view of a P-38J showing unique construction of the twin tailed aircraft. (US National Archives)

P-38J-15-LO, s/n 43-28338, "Bar Fly" of 93rd FS, 367th FG, 9th AF. Aircraft was assigned to 1st Lt. Earl D. Ody. Photo was taken on 7 October 1944 at Clastres, France. (US National Archives)

Bottom view of the P-38J, s/n 42-67543. (J. Kightly)

Above: P-38J registered as N3145X taking off. Undercarriage partially retracted. (J. Kightly)

Below: The same Lightning landing at Duxford during the annual Flying Legends show. (J. Kightly)

Above: Three destroyed P-38s scrapped at Shemya Island, Aleutian Islands. *(US National Archives)*

Below: Nose and port engine of 44-23314; a P-38J preserved by Planes of Fame, East in the USA. *(A. Lochte)*

MAIN FUSELAGE

P-38J-25-LO, s/n 44-23673, "Janet" of 393rd FS, 367th FG, 9th AF. Aircraft assigned to 1st Lt. William G. Norris. Photo was taken on 7 October 1944 at Clastres, France. Note details of the main fuselage panels. (US National Archives)

Right: Nose of the P-38, with major service access panel removed. Fuselage construction is visible. (J. Kightly)

Another wartime photo of the port side of the main fuselage, P-38J-10-LO, s/n 42-68176 this time. "Sky Cowboy" of 77th FS, 20th FG, 8th AF. Aircraft was lost on 12 July 1944 in a take-off accident. The pilot, Lt. Richard Robbins, was seriously injured. Note interesting kill and mission markings below the canopy.

Right: Port side of the main fuselage, rear underside view. Ladder in retracted position. Note that the red light is a modern addition, not fitted during WWII. (A. Lochte)

Above left: A main fuselage from the rear. Note, the pilot ladder in the centre in retracted position. Panel behind the rear of the canopy is access to a hand grip.
Above right: The pilot's ladder in down position.
Below: Bottom rear fuselage. Recognition lights and part of the retractable ladder are shown. (All photos A. Lochte)

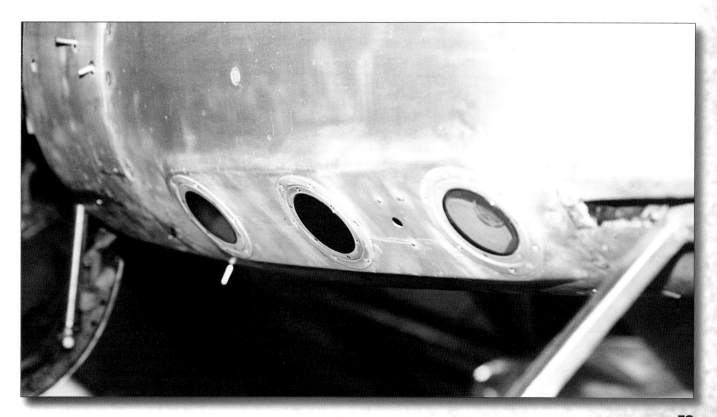

Port-rear part of the fuselage. Canopy and pilot's ladder in closed positions. (A. Lochte)

Bottom, rear part of the fuselage. Access ladder partially extended. (A. Lochte)

American pilot posing for a propaganda photo in Italy circa 1944. Hand held is clearly visible. (US National Archives)

TAIL BOOMS

Above: Wartime photo of two P-38s in England, late 1944. Inner and outer sides of the tail boom are visible showing panel lines and riveting. Both aircraft with D-Day stripes on the undersides of the booms.

Below: Destroyed P-38 in the Philippines, 1945. Cross section of the tail boom is clear.

Italy, 1944. Inner side of the port tail boom is visible. (Both photos US National Archives)

P-38J-15-LO s/n 43-28650, "Nellie Ann" of 27th FS 1st FG, Italy.

Above and right: details of the coolant radiator doors.

Bottom: Coolant radiator outlets from the rear. Flaps in fully open position.

(All A. Lochte)

Above: Two shots of the coolant radiator air intake.
Below: Coolant radiator air intake and supercharger air intakes.
(All photos A. Lochte)

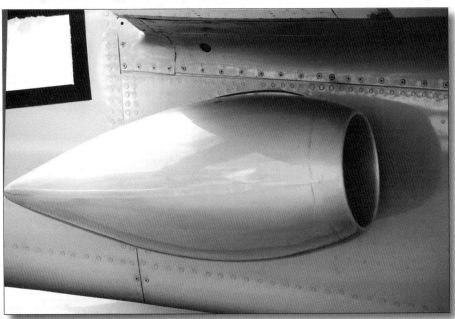

Supercharger air intakes. These are located just below the trailing edge of the outer wings on both booms.
(All photos A. Lochte.)

ENGINES

Somewhere in England, checking one of the Allison engines of the P-38 "No guts" usually flown by Cpl. Guy M. Purdy. All panels are removed showing engine details. (US National Archives)

Front view of the starboard engine.
(A. Lochte)

Details of the engine air intakes.
The outer two intakes shown in the top, left photo are for the oil cooler, while the centre intake (clearly shown in the right bottom photo) is for the intercooler.
In the right photos the exhaust shroud (large scoop) is visible and spark plug cooling intake.
(All photos A. Lochte.)

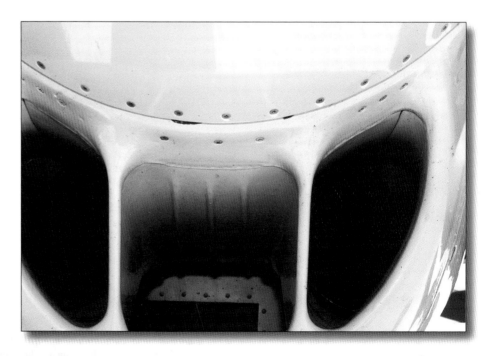

Above: Wartime photos showing engine maintenance on a P-39J. *(US National Archives)*

Colour photos: Photos of the starboard engine with cowling removed. Engine details and engine mount assembly are visible.
(All photos A. Lochte)

Above: Starboard side of the Allison engine with panels removed. Oil cooler is visible in the right bottom part of the engine. The black pipe is the plug spark cooling system.

Right: Details of the intercooler flap, beneath of the both engines. Flap in open position.

Bottom left: Rear of the engine cowling. Round cover on the door is access to the heater duct.

(All photos A. Lochte)

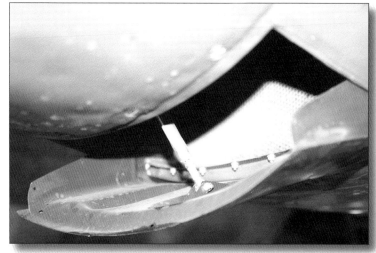

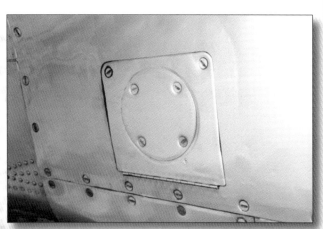

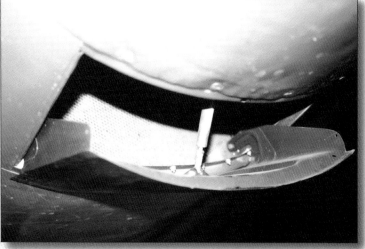

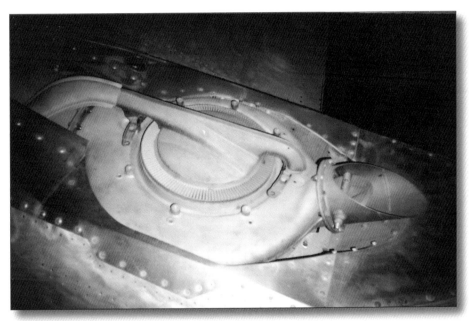

Above: Two photos of the air intakes for the turbocharger.
The outer intakes are for the turbocharger, and the centre one for cockpit heating.

Above, left: Overall view of the port turbocharger.

Left: Two photos of the General Electric Type B-33 supercharger.

(All photos A. Lochte)

Mechanics working on the exhaust-driven turbocharger. Upper panels are removed showing the boom construction. (US National Archives)

Right: G. E. type B-33 turbocharger, exhaust turbine. (A. Lochte)

Port engine, inner side. Note details of the spinner and propeller.
P-38L-1-LO, s/n 44-24008 "Little Red Head." Probably of "Lightning Provisional" group, 318th FG. Isley Field, Saipan, September 1945. (US National Archives)

Right: Close up view of the port engine cowling. Oil cooler in closed position. (A. Lochte)

Front view of the Allison V-1710 Model F engine. (P-38 Technical Manual)

Right: Complete General Electric Type B-33 turbo-supercharger installation. This supercharger was used on P-38J and P-38L aircraft. (Technical Manual)

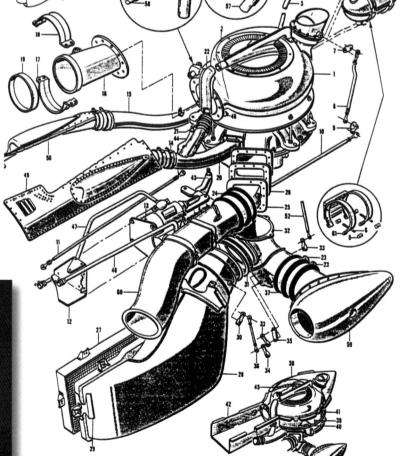

Below: Rear view of the B-33 supercharger. (A. Lochte)

CANOPY

Above: Port side of the canopy in closed position. (A. Lochte)
Below: Period photo of the starboard side of the canopy in open position. Note cockpit details. (US National Archives)

Above: *Port side of the canopy in open position. Details of the frame are visible.*
(US National Archives)

Colour photos: *Three photos of the P-38 canopy. Canopy with internal armour glass.*
(All A. Lochte)

Above: Rear, starboard part of the canopy. Radio compartment behind pilot's seat is clearly shown. (A. Lochte)

Below: P-38 cockpit canopy components. (Technical Manual)

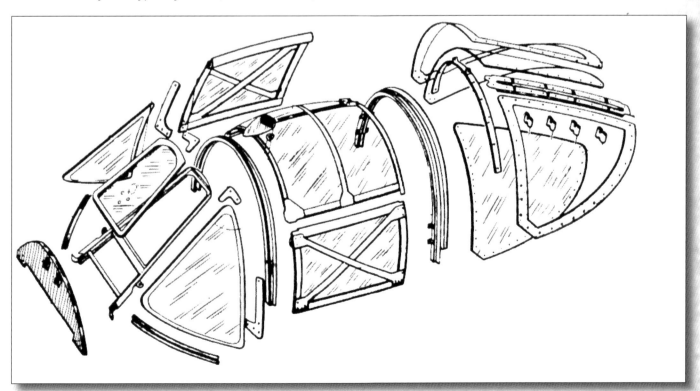

COCKPIT

The left, forward part of the cockpit. The throttle quadrant, landing gear controls and trim settings are visible.

Starboard cockpit panel. Controls of the SCR-522 radio system are visible. The handle visible just aft of the radio control is to raise and lower the side canopy panel.

Details of the unique P-38 control wheel. (All photos A. Lochte)

INSTRUMENT PANEL (P-38J-25)

1. Standby magnetic compass.
2. Suction gauge.
3. Clock.
4. Gyro horizon.
5. Manifold pressure gauges (left and right).
6. Tachometers (left and right).
7. Engine gauge right engine (oil temperature and pressure and fuel pressure).
8. Coolant temperature gauge.
9. Carburetor air temperature gauge.
10. BC-608 contractor (eliminated).
11. Generator switches.
12. Ammeters.
13. Compass correction cards.
14. Engine gauge left engine (oil temperature and pressure and fuel pressure).
15. Rate of climb indicator.
16. Bank and turn indicator.
17. Airspeed indicator.
18. Directional gyro.
19. Remote indicating compass.
20. Front (reserve) fuel tanks quantity gauge.
21. Rear (main) fuel tanks quantity gauge.
22. Hydraulic pressure gauge.
23. Altimeter.
24. Landing gear warning light.

25. Landing gear warning light test button.
26. Spare bulb.

MAIN SWITCH BOX (P-38L-5)
(below)

1. Ignition switches.
2. Oil dilution and engine primer switches.
3. Starter switch.
4. Engage switch.
5. Wing and tail position light switches.
6. Generator switches.
7. Landing light switch.
8. Gun heater switch.
9. Compass light switch.
10. Fluorescent light rheostat.
11. Voltmeter.
12. Ammeter.
13. Propeller feathering switches.
14. Oil cooler flap switches.
15. Battery switch.
16. Pitot heat switch.

17. Coolant flap override switches.
18. Intercooler flap switches.
19. Cockpit light rheostat.

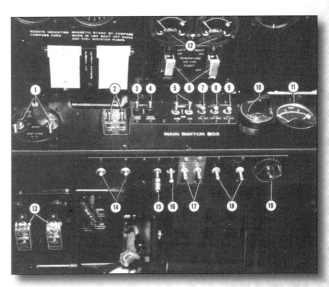

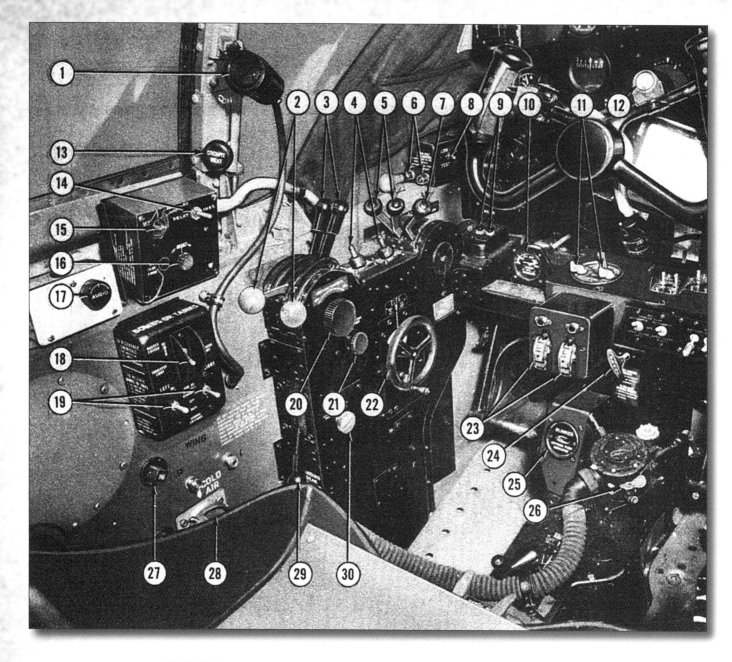

COCKPIT – LEFT SIDE (P-38L-5)

1. Spotlight(normal position).
2. Throttles.
3. Propeller governor controls.
4. Propeller selector switches.
5. Mixture controls.
6. Outer wing low level fuel warning lights.
7. Air filter control.
8. Outer wing low level fuel test switch.
9. Propeller circuit breaker buttons.
10. Oxygen pressure gauge.
11. Ignition switches.
12. Radio transmitter button.
13. Cockpit heat control.
14. Rocket arming switch.
15. Rocket selector switch.
16. Rocket reset knob.
17. Radio volume control.
18. Bomb-drop tank master switch.
19. Bomb-drop tank selector and arming switches.
20. Friction control.
21. Propeller lever vernier knob.
22. Elevator trim tab control.
23. Propeller feathering switches.
24. Parking brake handle.
25. Oxygen flow indicator.
26. Oxygen auto-mix lever.
27. Spotlight alternate position socket.
28. Cockpit ventilator control.
29. Landing gear control handle.
30. Landing gear control release knob.

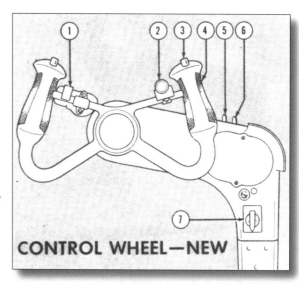

CONTROL WHEEL—NEW

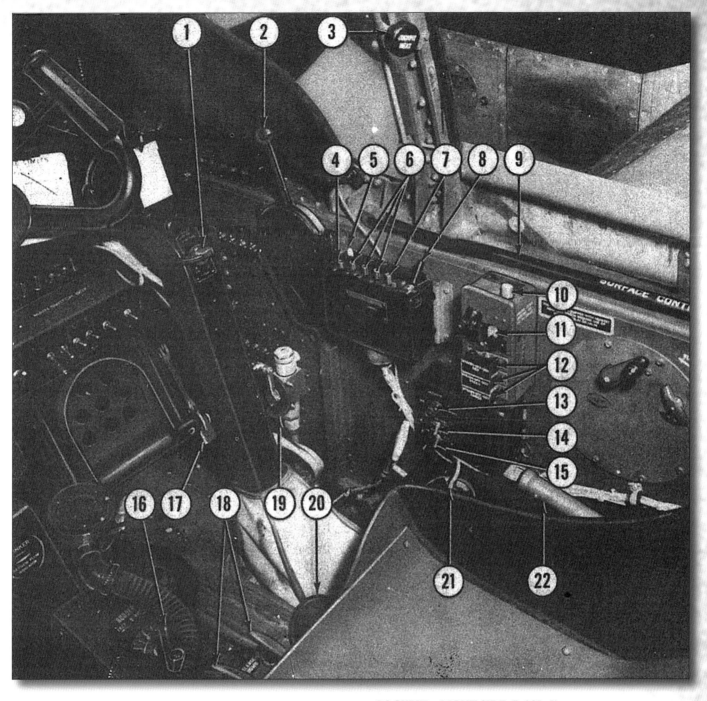

COCKPIT – RIGHT SIDE (P-38L-5)

1. Gunsight light rheostat.
2. Flap control lever.
3. Cockpit heat control.
4. VHF radio OFF push button.
5. Wing and tail position light switches.
6. Frequency selector push buttons.
7. Selector lock lever.
8. VHF radio control lever.
9. Surface controls lock (stowed).
10. Recognition light keying button.
11. Cockpit light.

12. Recognition light switches.
13. AN/APS-13 warning light rheostat.
14. AN/APS-13 test switch.
15. AN/APS-13 ON-OFF switch.
16. Rudder trim tab control.
17. Rudder pedal adjustment lever.
18. Manual bomb-drop tank release.
19. Aileron boost control lever.
20. Relief tube.
21. Low frequency range receiver.
22. Hydraulic hand pump.

CONTROL WHEEL
(opposite page – right)

1. Dive flap control.
2. Radio transmitter button.
3. Bomb-rocket release button.
4. Machine gun-cannon trigger button
(back of wheel).
5. Bomb-rocket selector switch.
6. Gun-camera selector switch.
7. Gunsight light rheostat.

14. TACHOMETER ADAPTER
15. RELEASE LEVER
16. CROSSHAIR RHEOSTAT
17. DRIFT SCALE
18. PDI BRUSH AND COIL
19. AUTOPILOT CLUTCH ENGAGING KNOB

Top: Sperry K-14 gun sight, which replaced the earlier gun sight. (US National Archives)

Above: *Drawing of the K-14 components. (Pilot Flight Manual)*

Right: *Instrument panel of the early P-38L – similar to P-38J. Lynn L-3 reflector gunsight is visible above the instrument panel. (US National Archives)*

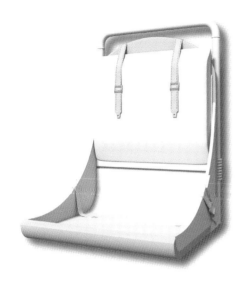

Pilot's seat, 3D drawing. (D. Grzywacz)

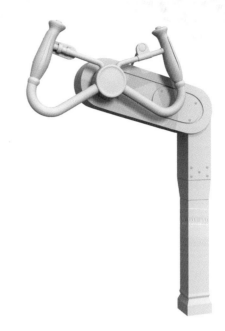

P-38 control column, 3D drawing. (D. Grzywacz)

Radio (274N type) checking during maintenance of the P-38L. (US National Archives)

TAIL

Above: Rear view of the P-38 showing the construction of the tail. (US National Archives)

Right: Almost a side on elevation of the starboard vertical stabiliser.
Note, that the antenna wire was added postwar in place of the navigation light.
Below: Starboard rudder navigation light. (photos A. Lochte)

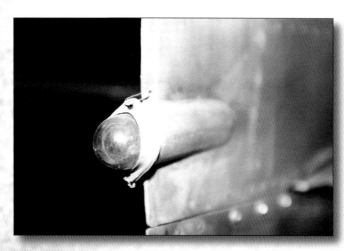

Left: P-38 tail unit.
Below: Photo of the vertical stabiliser, inner side.
(Photos A. Lochte)

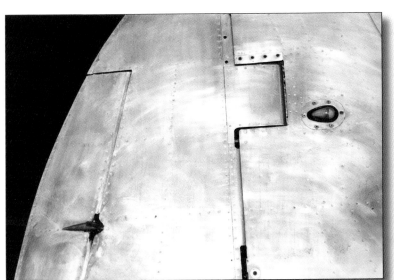

Above left: *Outside, starboard fin & rudder showing trim tab, navigation light and rudder mass balance.*

Left: *Details of the horizontal stabiliser and elevator mass balance.*
(A. Lochte)

105

Details of the horizontal stabiliser and elevator mass balance.

Bottom left: *Lower part of the starboard vertical stabiliser, inner side. (A. Lochte)*

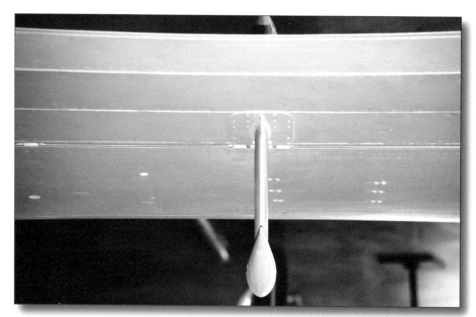

P-38L taking off, rear view. Note the landing flaps in the extended position. (US National Archives)

Tail components. Note the two piece rudder with upper and lower sections. The most lower part is a steel skid plate. (Technical Manual)

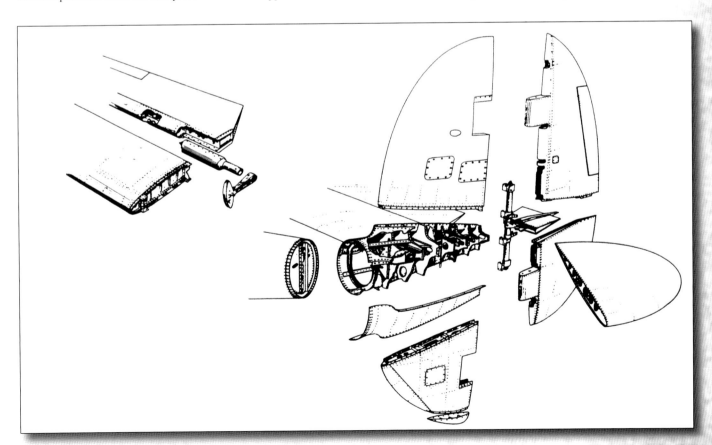

WING

Above: P-38L taking off. Flaps are in the fully extended position. (US National Archives)

Below: Bottom of the starboard wingtip. (A. Lochte)

Beautiful shot of P-38J-15-LO s/n 43-28650, "Sweet Sue/Nellie Ann", flown by Lt. Phillip E. Tovrea of 27th FS, 1st FG, Salsola, Italy, summer 1944, in flight. Underside wing panel lines are visible. Note underwing pylons.
(US National Archives)

Inner starboard wing. Pylon and drop tank are also visible.

Wing root fillet, port wing.

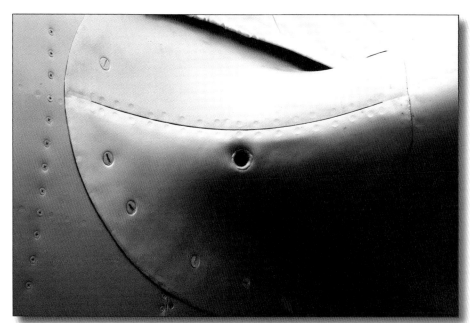

Upper side of the starboard inner wing. (All photos A. Lochte)

Port wingtip with pitot tube and navigation light.

Port wingtip, front view.

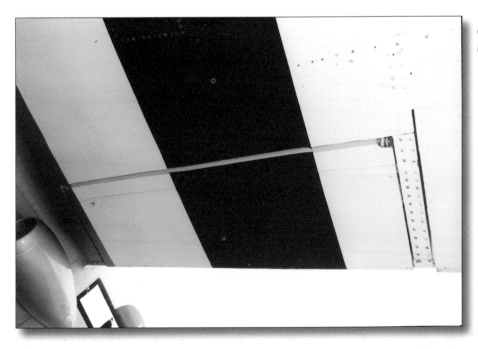

Outer Fowler flap – port wing. (All photos A. Lochte)

Photos of the dive flap.
Flap in open position in the middle photo.
(All photos A. Lochte)

Booster fuel pump is visible in front of the dive flap.

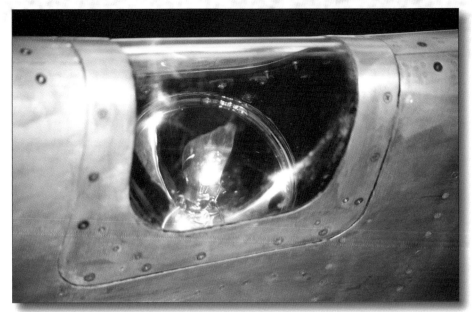

Landing lamp in the leading edge of the port wing.

Starboard drop tank pylon with gun camera.

Port drop tank pylon.
(All photos A. Lochte)

Right: Inside of the extended, port Fowler flap.

Below: Fowler flap in the full extended position, period photo. (US National Archives)

Below right: Flower flap construction (Technical Manual)

Bottom: Outer wing section joint construction. (Technical Manual)

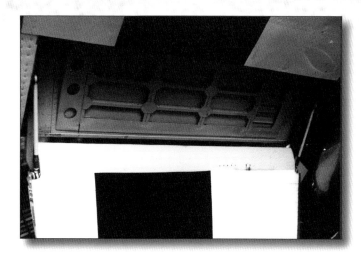

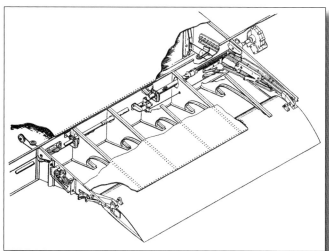

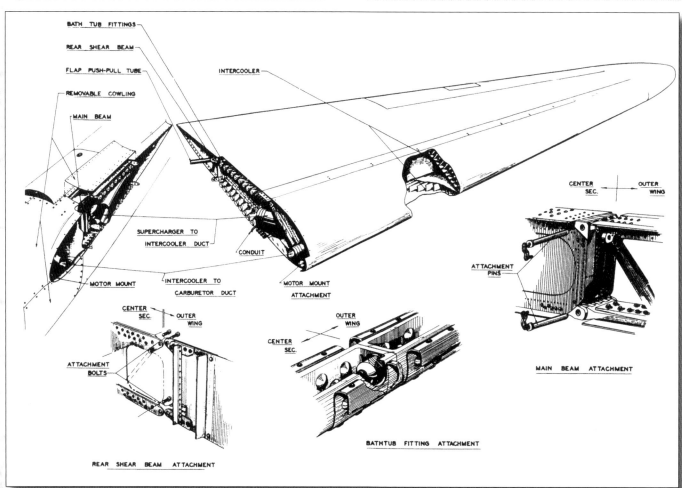

UNDERCARRIAGE

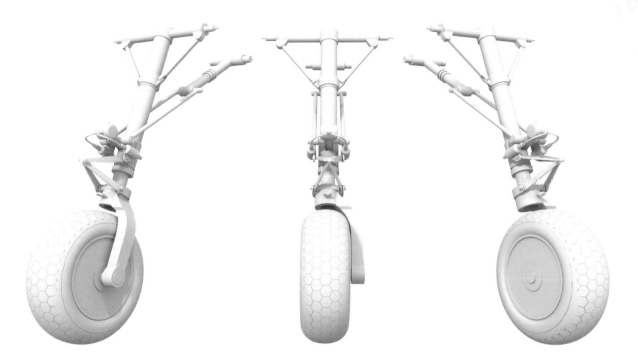

Above: 3D drawings of the nose gear undercarriage arrangement. (D. Grzywacz)

Below: Two photos of the nose undercarriage. Shimmy damper and torque link are visible. (A. Lochte)

Above, left: Nose gear showing drag struts, shimmy damper and torque link.

Above right: Nose undercarriage – front view.

Right and bottom right: Details of the nose gear wheel, port and starboard side.

Bottom, left: Details of the nose undercarriage door.

(All photos A. Lochte)

P-38L-5-LO, s/n 44-25686, "Li'l Margaret" of 1st FG, Italy, 1945.
Details of the nose undercarriage are visible. (US National Archives)

The nose undercarriage door, rear part.
Hinges and door actuation arm are visible.
(A. Lochte)

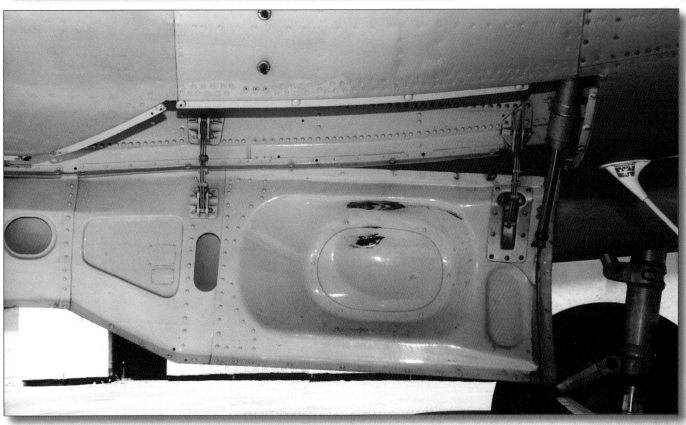

Above: Two photos of the nose undercarriage wheel well.
Below: Shimmy damper and torque link are shown in the photo. (All photos A. Lochte)

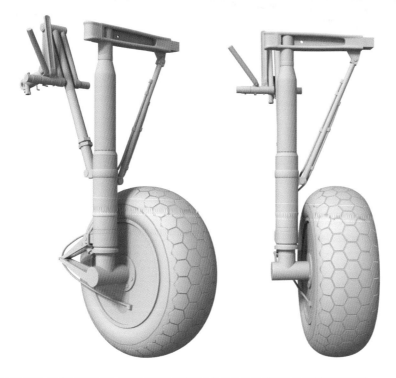

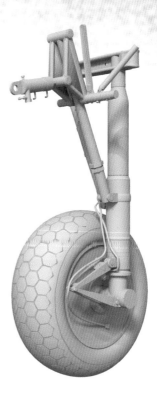

Above: 3D drawings of the main undercarriage arrangement. (A. Grzywacz)

Left: Two photos of the main undercarriage wheel.

Below: Front view of the main undercarriage. (All photos A. Lochte)

 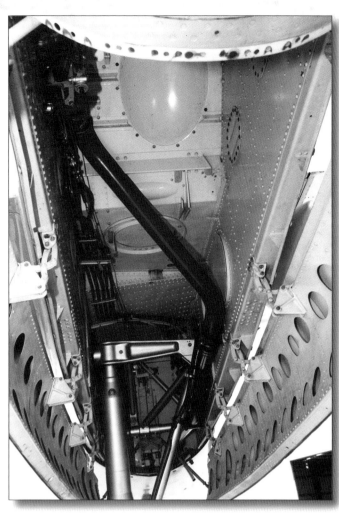

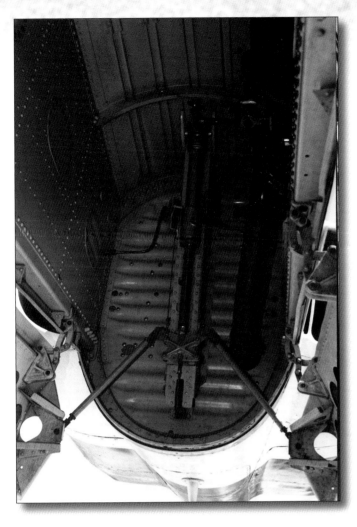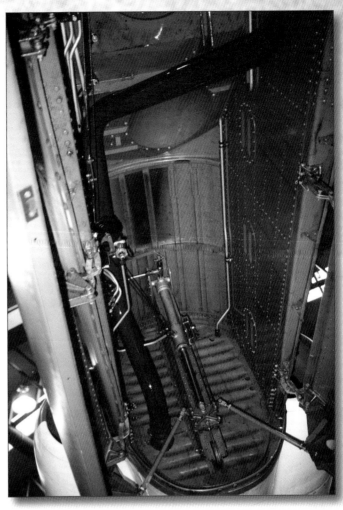

Above: Two photos of inside of the main gear wheel well, looking aft, port and starboard. Coolant pipe is blue this time.

Below: Details of the main undercarriage wheel well doors.

Opposite page:
Top left: *Details of the main undercarriage wheel.*

Top, right: *Main undercarriage wheel well looking in flight direction.*

Bottom, left: *Bottom of the starboard boom – gear doors open. Note the door hinges inside the wheel well.*
Also the oxygen access panel is visible, just aft the wheel well.

Bottom right: *Port main gear wheel well, looking forward. Long silver tube is coolant pipe.*
(All photos A. Lochte)

ARMAMENT

Above: *Rearming a P-38's machine guns. Note the unpainted areas of the internal surfaces of the armament bay. (US National Archives)*

Early version of the armament without the blast tubes.
(A. Lochte)

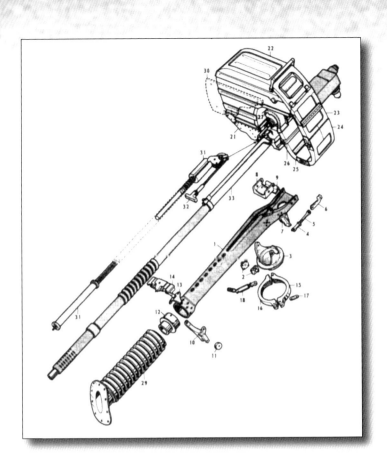

20 mm cannon with its mounts, ammunition container and recoil springs. (Technical Manual)

Below: Wartime photos of armament maintenance on a P-38L. Note that machine guns are fully loaded. (US National Archives)

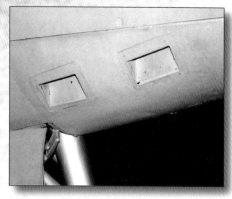

Above: Empty shell ejector chutes on starboard side. (R. Pęczkowski)

Right: Another wartime photo showing rearming of the machine guns. (US National Archives)

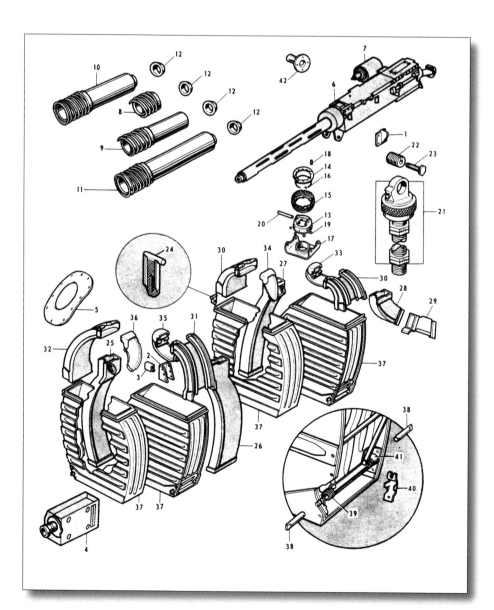

Machine gun arrangement. Drawing from Technical Manual

Three photos of the drop tanks and their pylons.
(US National Archives)

125

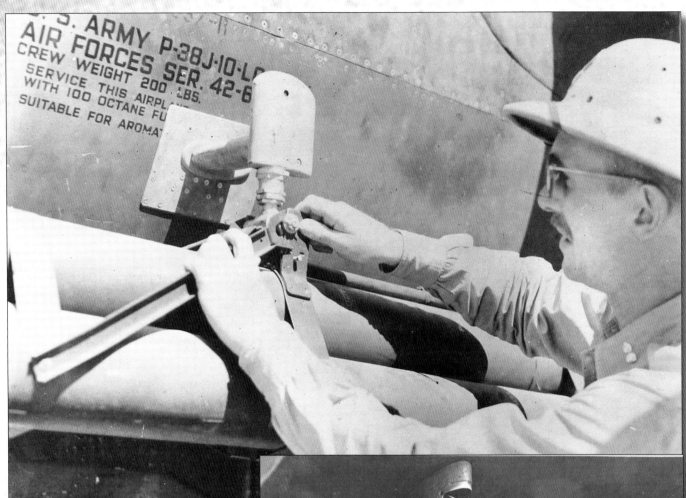

Above: P-38J-10-LO 42-6???? of 459th FS 80th FG. Usually piloted by Maj. Willard J. Webb. Aircraft has Bazooka launchers mounted to main fuselage sides. (US National Archives)

Right and below: Two photos of bomb loading. (US National Archives)

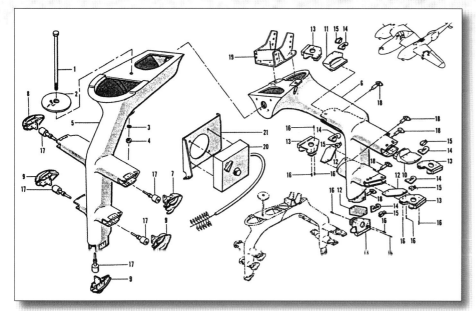

Christmas Tree launchers.
The cast light alloy supports are the main components. These were bolted together and attached under the wing.
The complete unit is shown at the bottom of the drawings. (Technical Manual)

Photo of the Christmas Tree launcher under the port wing.
(US National Archives)

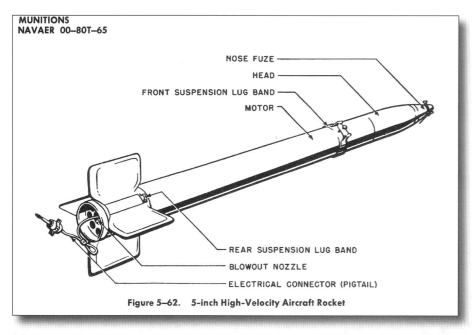

MUNITIONS
NAVAER 00–80T–65

NOSE FUZE
HEAD
FRONT SUSPENSION LUG BAND
MOTOR

REAR SUSPENSION LUG BAND
BLOWOUT NOZZLE
ELECTRICAL CONNECTOR (PIGTAIL)

Figure 5–62.   5-inch High-Velocity Aircraft Rocket

Drawing of the HVAR rocket. (Technical Manual)

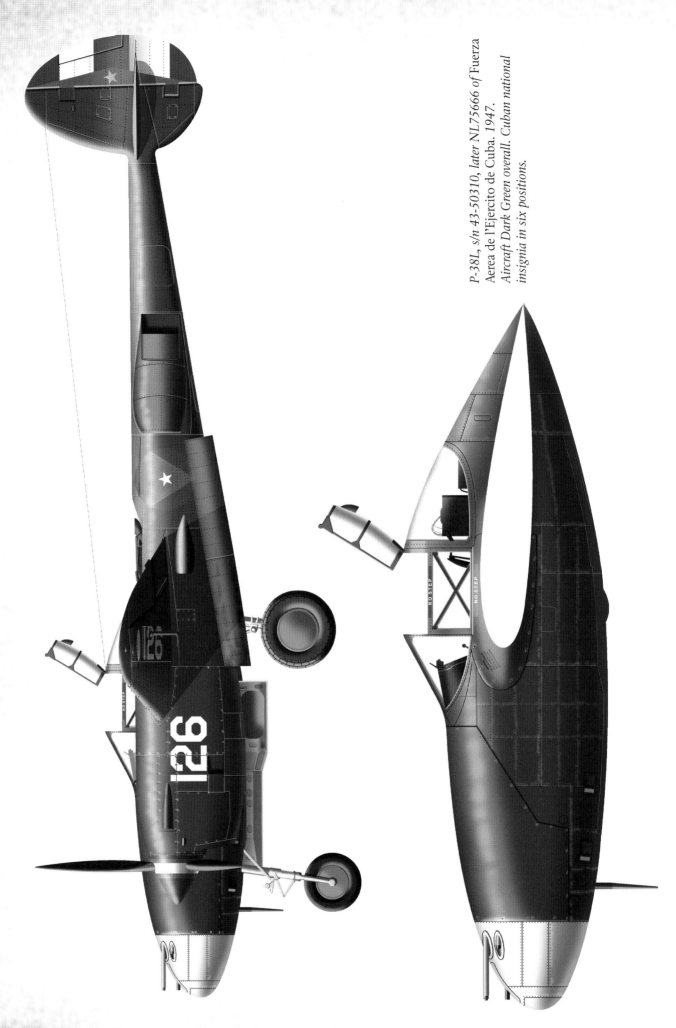

P-38L, s/n 43-50310, later NL75666 of Fuerza Aerea de l'Ejercito de Cuba. 1947. Aircraft Dark Green overall. Cuban national insignia in six positions.